Indian Herbs for Touch Therapy, Tattooing and Adornments

Indian Herbs for Touch Therapy, Tattooing and Adornments

S.K. Sood
Vipula Sharma
T.N. Lakhanpal
Romita Sharma
Suresh Kumar

Department of Biosciences
Himachal Pradesh University
Shimla – 171 005, H.P.

2011

Daya Publishing House®
A Division of
Astral International Pvt. Ltd.
New Delhi – 110 002

ISBN 9789351240730

Published by : **Daya Publishing House®**
A Division of
Astral International Pvt. Ltd.
– ISO 9001:2008 Certified Company –
4760-61/23, Ansari Road, Darya Ganj
New Delhi-110 002
Ph. 011-43549197, 23278134
E-mail: info@astralint.com
Website: www.astralint.com

Laser Typesetting : **Classic Computer Services**
Delhi - 110 035

Printed at : **Salasar Imaging Systems**
Delhi - 110 035

PRINTED IN INDIA

Abbreviations

α	Alpha
&	and
A.P.	Andhra Pradesh
β	Beta
cm	Centimetre
δ	Delta
D	Dextro rotatory
E.	East / Eastern
Eng.	English
et al.	*et alii* (and other authors)
etc.	*et cetera* (so on)
γ	Gamma
g	Gram
H.	Hindi
H.P.	Himachal Pradesh
hrs	Hours
i.e	*id est.* (that is)
J. & K.	Jammu and Kashmir
Kg	Kilogram
M.P.	Madhya Pradesh
m	Meter
Me	Methyl
N.	North/Northern
N	Nitrogen
N.E.	North East
N.W.	North West
P	Para
%	Per cent
PS	Polysaccharide
Ψ	Psi
Sans.	Sanskrit
S.	South/ Southern
S.E.	South East
S.W.	South West
sp	Species
spp	Species
Ssp	Subspecies
Syn.	Synonymous
U.P.	Uttar Pradesh
var.	Variety
vit.	Vitamin
Viz.	*Vide licet* (namely)
Vol. (s)	Volume (s)
W.	West / Western
W.B.	West Bengal

Preface

Nearly all cultures from ancient times to present day have used plants for their sustenance. Ironically, most of the age-old plant based tribal traditional practices developed over years of quest through trials and errors are vanishing due to varied biotic pressures and changes in sustenance economy, and needs to be popularized with the present and future generations for which the need of the hour is to develop sound databases on their ethnobotanical prudence. Based on this rationale, the preparation of this multidisciplinary compendium bringing together for the first time our large mass of scattered ethnobotanical data from disparate interdisciplinary sources, has been necessitated with all earnest efforts to provide up-to-date information for discernible users on botanical names, synonyms, families, English, Hindi and Sanskrit names, distribution, reproductive cycle, part(s) used, folk use(s), active constituents of 440 species belonging to 347 genera under 121 families of Indian herbs used for touch therapy, tattooing and adornments of which 1 species (1 genus) belongs to Algae, 3 species (3 genera) to Fungi, 11 species (9 genera) to Pteridophytes, 78 species (56 genera) to Monocots and the remaining 347 species (275 genera) to Dicots with ratio of species, genera and families between Monocots and Dicots as 1: 7.75, 1: 4.91, 1: 4.39, respectively. For amply illustrating the textual matter and making the reading absorbing, 1 pie chart, 7 histograms and 31 photographs have been included. For ready reference to contents, 20 useful appendices have also been provided. The authors believe that this compendium with its rich mine of information on the subject will go a long way in bridging the gap between the generators of information on the subject and the commercial houses. As a source of ready reckoner on the subject, it is indispensable and no institutional library should be without it.

Well deserved appreciation is due to Prof. A.D.N. Bajpayee, Vice-Chancellor, Himachal Pradesh University, Shimla, Prof. H.S. Banyal, Prof. D.C. Kalia, Prof. D.C. Gautam and Prof. V.K. Mattu for their well wishes and incessant encouragement; to

Mr Ved, Ms Seema, Ms Suman, Ms Chandrika and Ms Parul for their ungrudging help in critically reading the various drafts of this manuscript, and to Mr Bal Krishan for his help in various ways. Our sincere thanks are also due to M/S Daya Publishing House, Delhi for bringing out this book in an attractive form. Thanks are also due to authors of books, papers and figures listed in the treatise. Finally, we would like to thank the members of our families for bearing the opportunity cost.

Suggestions for further improvement of the treatise will always be welcomed and entertained whole heartedly.

S.K. Sood

Vipula Sharma

T.N. Lakhanpal

Romita Sharma

Suresh Kumar

Contents

Introduction

Indigenous People (= Rural Communities/Tribals/Ethnic Societies/Traditional Societies)

India with over 550 natives or tribal communities under 227 ethnic groups covering 5000 villages has been rightly described as a treasure trove of ethnobotanical knowledge due to presence of multi-ethnic groups of ancient lineage and very diverse vegetation (Anonymous, 1994). Our age-old culture and traditions with obligation to natural entities have further contributed to the richness of heritage (Kosambi, 1962). Indigenous societies all over the world in different geographical regions have discovered multiple uses of natural resources around them in the form of traditional knowledge (Jain, 2004). There is no gain saying the fact that traditional knowledge survives usually among the indigenous and local communities as they maintain a balanced ecological rhythm in their surroundings. Traditionally, local communities worldwide are extremely knowledgeable about the local plant resources on which they are so intimately and immediately dependent. In fact, these societies are human conservatories which can't be duplicated by application of science and technology. Unfortunately, much of the accumulated extraordinarily abundant knowledge on plants which have been acquired due to their long-term practices and handed down usually orally, form one generation to another, is dwindling because of the loss of their main culture and changes in sustenance economy (Hamilton, 1955). Therefore, the need of hour is to collect and inventorise the traditional knowledge and their resource management practices, before these are lost forever to posterity.

Touch Therapy, Tattooing and Adornments

Touch therapy, a method of treatment of ailments commonly practised by tribal societies, involves touching the body of the ailing person or domestic animals with locally available plant species for the prevention and/or cure of the diseases. In touch therapy, plant part/s are usually tied to the body of patient/ailing animal or hung around the neck with a thread by the medicinemen for the desired effect. Mainly, the therapy is practised for curing various ailments in children and women. These are: jaundice, fever, hemicrania, hoarse cough, boils, goitre, epilepsy, urticaria, abortion, post-delivery treatment, headache, rheumatism, sunstroke and inducing labour pain (Hembrom, 1996; Punjani, 2002). This method of treatment has been suggested as more a magic than treatment and called as magical cure of diseases by Jain (1963). Nevertheless, this type of administration of the medicine attracts populace both from the rural and urban areas (Sen and Behera, 2007).

The practice of tattooing the body is very old. The explorer James Cook introduced the word to English speakers in his account of a voyage around the world from 1768 to 1771. Tattooing involves making coloured patterns or pictures on part of a person's body by pricking the skin and putting in dyes. Generally, tattooing is done in the childhood and brought about by tribals using either pointed spines of some plants or copper needles to decorate their body parts like arms, breasts, wrists, face, belly and legs (Jain, 1995). Tattooing is done to have relief against some specific ailments.

The term adornment in the widest possible sense refers to decoration and it includes 2 classes: ornaments (beads and pendants) and cosmetics (pigments or oils applied directly to the surface of hair, skin, teeth or nails) (Francis, 1984). Primitive peoples in various parts of the world use plants for body paints, adornment, living stockades to protect themselves from each other and from wild animals, poisons, chewing, stimulants, medcines and ritual purposes (Mehra *et al.*, 1975). India's long literary tradition furnishes widespread use of plants for adornment. The great Indian poet Kalidasa (4th century AD) recorded several flowers used in this manner (Mehra *et al.*, 1975). Furer-Haimendorf (1943) noted that the Chenchu women decorated their bodies with any attractive bloom at hand. Also, plants are commonly adorn to indicate an individual's religious affiliation, *e.g.*: beads of *Elaeocarpus ganitrus* worn by followers of Shiva; necklaces made from the woody parts of *Ocimum sanctum* by Vishnu worshippers and wooden beads of *Aegle marmelos* worn by Sudras to show they are Hindus (Francis, 1984; Sood *et al.*, 2005). Plant fibers are commonly used to string beads (Vidyarthi, 1975). Undeniably, Indian women use several species of plants for making a wide variety of ornaments for personal adornment and for beautifying their homes.

Current Status on Herbs for Touch Therapy, Tattooing and Adornments

India is known for its ethnobotanic endeavours right from the time of the Vedas and Samhitas. But the real credit to explore systematically the plant wealth of India goes to Heinrich Van Rheede-the Governer of Dutch possession in Malabar in 1667. Publication of Watt's (1896) '*Dictionary of the Economic Products of India*' was a landmark in modern ethnobotany. '*Glimpses of Indian Ethnobotany*' by Jain (1981) is the first book dealing with Indian Ethnobotany, and the first attempt to publish annotated bibliography of ethnobotanical works describing over 2000 Indian and foreign references on various aspects of ethnobotany goes to Jain *et al.* (1984). Workers like Mudgal (1987) and Chandra (1990) compiled the diverse ethnobotanical literature existing in various languages of the world. Manilal (1989) commented on the linkages of ethnobotany with other sciences and disciplines. Another annotated publication, '*Dictionary of Ethnoveterinary Plants of India*' (Jain and Srivastava, 1999) describes briefly 836 plants for animal diseases and contributed significantly on veterinary phytomedicines. In recent years, a graphic review of Indian ethnobotanical literature with analysis highlighting possible directions for future work (Jain and Srivastava, 2001) and also a detailed one listing 1,600 references of which over 1,250 relate strictly to ethnobotanical work or very closely allied themes on India and the others on concept, global evolution of the discipline, and methodology of new areas from work of some outstanding foreign ethnobotanists have also enriched our understanding of the subject (Jain, 2002).

Survey of above literature reveals that available information on the subject and related aspects is not only scanty but scattered for India in general, *viz.*, *adornments* (Mookerji, 1930; Sastry, 1966; Mehra *et al.*, 1975; Malandhra, 1979; Francis, 1981, 1984, 1997; Vihari, 1995; Nazarudeen *et al.*, 1996; Rajendran and Aswal, 2000; Jha and Goel, 2004; Singh and Narain, 2008), *contact therapy* (Modi, 1893; Hembrom, 1996; Punjani, 2002; Bhatt *et al.*, 2003; Sen and Behera, 2007; Jadhav, 2008), *tattooing* (Jain, 1995; Sharma, 2000) and socio-magico-religious and sacred plants (Jain, 1963, 1991; Banerjee, 1974; Sebastian and Bhandari, 1984; Upadhyaya *et al.*, 1997; Bhatt *et al.*, 2003; Chakraborty and Bhattacharya, 2004; Hajra, 2004; Khumbongmayum *et al.*, 2005; Negi, 2005; Sood *et al.*, 2005). In the absence of any comprehensive and multidisciplinary reference work so far, an attempt has been made presently to consolidate our traditional knowledge on this aspect from disparate sources and put it in proper perspective for the welfare of mankind.

Enumeration

For collecting data of interest on the ethnic herbal resources of India for touch therapy, tattooing and adornments, the research methodology as outlined by Schultes (1962), Jain (1964, 1973) and Parabia and Reddy (2002) was adopted to scrutinize the old literature (herbals, books, floras, compendia, materia medica, monographs, ancient treatises, research publications; Atkinson, 1882; Watt, 1889–1896; Anonymous, 1948–1976; Krishnamurthy, 1953; Sundararaj, 1959; Nadkarni, 1954; Agarwal, 1986; Ambasta, 1986; Nayar *et al.*, 1989; Maheshwari, 1996, 2000; Singh, 1996; Kumar, 1999; etc.). The information collected was compiled and collated in a set format designed by modifying that of Saklani and Jain (1994). For other aspects of information on synonyms, English, Hindi and Sanskrit names, distribution, reproductive cycle, active constituents and trade name(s), the works of Hooker (1872–1892), Kirtikar and Basu (1984), Anonymous (1948–1976), Chopra *et al.* (1956, 1980), Rastogi and Mehrotra (1990, 1991, 1993, 1995 a, b), Asolkar *et al.* (1992), Bhattacharjee (1998), Sood *et al.* (2001) and Sood and Thakur (2004), have been consulted and referred. Besides, data gathered from own field ethnobotanical investigation was also included, and marked with asterisk in the text.

The present database includes information pertaining to botanical names of the plants arranged in alphabetic order along with their synonyms, if available and respective families. Other details given are English, Hindi and Sanskrit names, distribution, reproductive cycle, part(s) used, folk use(s), active constituent(s), relevant references and trade name(s), if any. For convenience of reference and understanding, the text has been illustrated with the help of 31 photographs, 7 histograms and 1 pie–chart. Besides, information on total number of Indian genera and species for adornments, touch therapy and tattooing under various divisions of plant kingdom (*Appendix* I), families of adornments, touch therapy and tattooing of ethnic India (*Appendix* II), plants used for adornment (*Appendix* III), plants used for touch therapy (*Appendix* IV), plants used for tattooing (*Appendix* V), plants commonly used for adornments, touch therapy and tattooing (*Appendix* VI), predominant families (*Appendix* VII), predominant genera of adornment (*Appendix* VIII), predominant genera of touch therapy (*Appendix* IX), predominant genera of tattooing (*Appendix* X), plant parts used (*Appendix* XI), plant parts used for adornments (*Appendix* XII), plant parts used for touch therapy (*Appendix* XIII), plant parts used for tattooing (*Appendix* XIV), predominant dicot families (*Appendix* XV), predominant monocot families (*Appendix* XVI), botanical names (*Appendix* XVII), English names (*Appendix* XVIII), Hindi names (*Appendix* XIX), Sanskrit names (*Appendix* XX) have been included as appendices.

Indian Herbs for Touch Therapy, Tattooing and Adornments

Abelmoschus moschatus Medic.

Syn: *Hibiscus abelmoschus* Linn.; *H. chinensis* Wall.; *H. flavescens* Cav.; *H. ricinifolius* Wall.; *H. sagittifolius* Kurz.; *H. spathaceus* Wall.

Family: Malvaceae

Eng., H. and Sans. Names	:	*Eng.*: Ambrette plant, Musk mallow, Musk seed. *H.*: Kasturidana, Latkasturi, Mushkdana. *Sans.*: Gandapura, Kasturi, Latika, Latakasturika.
Distribution	:	Cultivated in hotter parts of the country.
Part Used	:	Roots.
Reproductive Cycle	:	June–January.
Folk Use	:	Root pieces strung and worn as garland to cure jaundice.
Trade	:	Mushkdana.
References	:	Ambasta (1986), Dhiman (2003), Khan and Khanum (2005), Singh and Pandey (1998), Sood and Sudershna (2008).

Abroma augusta Linn.

Syn: *A. angulata* Lam.; *A. fastuosa* Gaertn.

Family: Sterculiaceae

Eng., H. and Sans. Names	:	*Eng.*: Devils cotton, Perennial Indian hemp. *H.*: Kumal, Ulatkambal. *Sans.*: Peevaree, Ultkambal, Yoshinee.
Distribution	:	Cultivated throughout the hotter parts of India.
Part Used	:	Stem bark.
Reproductive Cycle	:	October–January.
Folk Use	:	A small piece of stem bark bandaged to the rectocele in children for 2-3 days to give complete relief.
Active Constituents	:	Friedelin and β-sitosterol characterized in stem bark.
References	:	Ambasta (1986), Asolkar *et al.* (1992), Kirtikar and Basu (1984), Prajapati *et al.* (2006), Rawat and Chowdhury (1998).

Abrus precatorius **Linn.**

Syn: *A. minor* Desv.; *A. pauciflorus* Desv.; *Glycine abrus* Linn.

Family: Fabaceae

Eng., H. and Sans. Names :	*Eng.*: Crab's eye, Indian liquorice, Jery quity, Paper noster pea, Rosary pea, Weather plant. *H.*: Chirmiti, Gnunghachi, Gunj, Karjini, Rati. *Sans.*: Bhilabhushana, Gunja, Krishnala, Raktika, Tulabija.
Distribution :	Throughout N.E. region.
Parts Used :	Stem, Root, Seeds.
Reproductive Cycle :	September–January.
Folk Use :	Stem made into ring and held in right hand to exorcize whereas its pieces tied on the left hand on Sunday/ Tuesday to cure fits. Tribals of Banswara district keep its roots for a few minutes in the navel of a pregnant woman for relief in delivery pains and to facilitate child birth. Garland made of seeds believed to be effective in nervous complaints like sciatica, stiffness of shoulders, joints and paralysis. Seeds used for making ear rings, rosaries worn as an ornament on the body parts by nomads, whereas its amulet believed to ward off evil spirits from the children by Bhils and Garasias in Rajasthan.
Active Constituents :	Active ingredients are abrin, a toxalbumin similar to ricin from castor seeds, a glucoside abralin, an alkaloid abrine and abarnin (seeds) precol, abrol (roots) galactose, arabinose and xylose (stem).
Trade :	Ratti or Gunchi.
References :	Chauhan (1999), Jha and Goel (2004), Kshirsagar and Singh (2007), Mehra *et al.* (1975), Nazarudeen *et al.* (1996), Radhakrishnan *et al.* (2000), Singh and Narain (2008), Singh and Pandey (1998), Sood *et al.* (2005), Uniyal (2007).

Abutilon indicum **(Linn.) Sweet.**

Syn: *A. asiaticum* Linn.; *A. guineense* Schum.;
***Sida guineensis* Schum.; *S. indica* Linn.**

Family: Malvaceae

Eng., H. and Sans. Names :	*Eng.*: Chinese bell flower, Country mallow, Flowering maples. *H.*: Jhampi, Kandhi, Kanghi, Potari, Tepari, Tutri.

		Sans.: Atibala, Balika, Balya, Chanta, Kankati, Shita, Vikankala.
Distribution	:	Throughout the tropics and sub-tropics; cultivated as well occurring naturally as a common weed of roadsides.
Part Used	:	Seeds.
Reproductive Cycle	:	May–January.
Folk Use	:	Seeds used as ear pendant; also tied over the joints tightly with a piece of cloth against joint pains.
Active Constituents	:	Alantolactone and isolantolactone are the chief constituents of seeds.
References	:	Ambasta (1986), Jain (1991), Jain and Mudgal (1999), Katewa and Sharma (1998), Seetharami Reddi *et al.* (2005), Sood and Sudershna (2008).

***Acacia catechu* Willd.**

Syn: *A. catechuchoides* Wall.; *A. polycantha* Willd.; *A. sundra* Bedd.; *A. wallichiana* DC.; *Mimosa catech* Roxb.; *M. catechuchoids* Roxb.

Family: Fabaceae

Eng., H. and Sans. Names	:	*Eng.*: Black cutch, Catechu, Cutch tree. *H.*: Katha, Khadir, Khadra, Khair. *Sans.*: Dantadhavana, Jivumanglaya, Khadira.
Distribution	:	Panjab, M.P., U.P., A.P. and Bihar ascending to 1,500m.
Part Used	:	Wood.
Reproductive Cycle	:	May–July.
Folk Use	:	Wood cut into the plow-shaped amulet for adornment.
Active Constituents	:	Wood contains catechin and catechutanic acid.
References	:	Ambasta (1986), Chauhan (1999), Harborne and Baxter (2001), Shende (1949), Sood *et al.* (2005).

***Acacia ferruginea* DC.**

Family: Fabaceae

Sans. Names	:	*Sans.*: Buahmashalya, Dvijapriya, Kadara, Kantadhya, Somasara, Somavriksha.
Distribution	:	Throughout India, often planted in gardens.
Part Used	:	Flowers.
Folk Use	:	Used as adornments.
References	:	Dhiman (2003), Gogoi and Borthakur (1991).

Acacia leucophloea **(Roxb.) Willd.**

Syn: *A. alba* Willd.; *A. arcuata* Decaisne.; *Mimosa alba* Rottl.; *M. leucophloea* Roxb.

Family: Fabaceae

Eng., H. and Sans. Names :	*Eng.*: Distiller's acacia, White Babool. *H.*: Arinj, Jhind, Karir, Nimbar, Reru, Ringa, Safed-Babool, Safed kikar. *Sans.*: Arimeda, Hemagaura, Kinkirata, Pitaka, Shvetabarbura, Somavalka.
Distribution :	Panjab, Rajputana, C. India.
Parts Used :	Stem bark. Root.
Reproductive Cycle :	August–December.
Folk Uses :	An amulet comprising its root, seeds of *Abrus precatorious* and bark of *Dichrostachys cinerea* believed to protect adults against evil eyes; also tied on arm to cure fever. Stem bark bandaged around dislocated bones of animals.
Active Constituents :	Stem bark characterized by the occurrence of n-hexacosanol, β-amysin and β-sitosterol whereas root possesses pimarene diterpenoid, leucophleol, leucophleoxol and betulinic acid-3-0-β-maltoside.
Trade :	Safed–kikar.
References :	Ambasta (1986), Asolkar *et al.* (1992), Bhatt *et al.* (2003), Khan and Khanum (2005), Rastogi and Mehortra (1980), Sikarwar (1996), Sood *et al.* (2005).

Acacia nilotica **(Linn.) Del. ssp. *indica* (Benth.) Brenan.**

Syn: *A. arabica* (Lamk.) Willd. var. *indica* Benth.; *A. arabica* Sensu Baker.; *A. vera* Willd.; *Mimosa arabica* Lam.

Family: Fabaceae

Eng., H. and Sans. Names :	*Eng.*: Black Babool, Indian gum arabic tree. *H.*: Babla, Babul, Babur, Kikar. *Sans.*: Ajabaksha, Babbuka, Babbula, Dirghakantaka, Goshringa, Kantalu, Svarnapushpa, Yugalaksha.
Distribution :	Common in warmer areas in Nahan, Nalagarh, Una, Bilaspur, Hamirpur and Kangra also in Mandi, Chamba, Kunihar.
Part Used :	Root.
Reproductive Cycle :	July–Sept and April–June.

Folk Use : Pieces of root tied on left hand with the help of "luccha" (sacred thread) to cure snake-bite.

Active Constituents : Roots contain octacosanol, beutlin, beta-amysin, beta-sitosterol, hentriacontane, lapachol, n-hentriacontanol, β-sistosterol and paulownin.

Trade : Indian Babul.

References : Ambasta (1986), Asolkar *et al.* (1992), Chauhan (1999), Dhiman (2007), Jadhav (2008), Khan and Khanum (2005), Rastogi and Mehrotra (1980-1984), Sharma (2003).

Acacia pennata (Linn.) Willd.

Syn: *A. megaladena* Desv.; *A. pennata* Datz. and Gib.; *Mimosa pennata* Roxb.; *M. tora* Roxb.

Family: Fabaceae

Eng., H. and Sans. Names : *Eng.*: Plume acacia.

H.: Aila, Biswul.

Sans.: Ari.

Distribution : Throughout India.

Parts Used : Twigs, Roots.

Reproductive Cycle : Flowering in rainy season and fruiting in cold season

Folk Use : Twigs and roots cut into beads for adornment.

References : Ambasta (1986), Anonymous (1948), Dastur (1964), Varghese (1996).

Achillea millefolium Linn.

Syn: *A. lanulosa* Nutt.

Family: Asteraceae.

Eng. and H. Names : *Eng.*: Arrow root, Blood wort, Camil, Carpernter-grass, Devils Nettle, Devils Plaything, Eerie, Green, Milfoil, Nosebleed, Yarrow.

H.: Gandana.

Distribution : W. Himalaya (Kashmir–Kumaon).

Part Used : Flowers.

Reproductive Cycle : July–Sept.

Folk Use : Used as ear top.

Active Constituents : Flowers possess azulene, achilloic acid, achilleine, a bitter principle, ivain and tannin.

Trade : Millfoil.

References : Ambasta (1986), Chauhan (1999), Jain (1991), Jain *et al.* (1991), Kirtikar and Basu (1984), Sharma (2003).

Achyranthes aspera **Linn.***

Syn: *A. bidentata* Blume.; *A. lanceolata* Wall.; *A. wightiana* Wall.; *Digera muricata* Mart.

Family: Amaranthaceae

Eng., H. and Sans. Names : *Eng.*: Prickly chaff flower plant.

H.: Apang, Chichra, Chirchira, Latjira, Puthkanda.

Sans.: Adhoganta, Aghata, Apamarga, Malakanta, Mayuraka.

Distribution : Throughout India upto 1,300m on roadsides, wastelands and pastures.

Parts Used : Leaves. Root. Stem.

Reproductive Cycle : April–September.

Folk Uses : A necklace made of small pieces of stem worn around the neck of women to facilitate child birth. Leaves tied on horns for wiping wound maggots by Soligas.Roots collected at the time of solar eclipse tied to the arm or hung around the neck for relief against all types of fever, the ill effect of Mercury planet and goitre; also tied on Sunday/Tuesday on pelvic region to facilitate delivery. For warding off ghosts and curing fever, its roots, twig and burnt leaves worn as an amulet on men's and women's arm. Pieces of root also tied to the ear for remedy in malarial fever and when tied on the waist cure dysentery. In Bastar district roots briefly worn or kept under the pillow of pregnant lady to accomplish easy delivery.

Active Constituents : Roots contain ecdysterone, ecdysone, inokosterone whereas stem contain pentatriacontane, 6-pentatriacontanone, hexatriacontane and tritriacontane.

Trade : Prickles chaff.

References : Asolkar *et al.* (1992), Banerjee (1974), Bhatt *et al.* (2003), Dhiman (2007), Hembrom (1996), Jain and Potole (2001), Katewa *et al.* (2001), Kshirsagar and Singh (2007), Mishra (2008), Punjani (2002), Rastogi and Mehrotra (1970-1979), Saini (1996), Saklani and Jain (1994), Sen and Behera (2007), Singh (2000), Sinha and Sinha (2001), Sood *et al.* (2005), Subramaniam (2000).

***Acorus calamus* Linn.**

Syn: *A. belangeii* Schott.; *A. calamus* var. *verus* Linn.; *A. casia* Bertol.; *A. griffithii* Schott.

Family: Araceae

Eng., H. and Sans. Names : *Eng.*: Bastard Water flag, Calamus, Myrtle grass, Sweet flag, Sweet myrtle.

H.: Bach, Bhoja, gorbach, Igir, Jatila, Safed Bach, Vacha.

Sans.: Bhadra, Bhutanashini, Galani, Ugragantha, Vacha, Vijaya.

Distribution : Himalaya. N. E. India. Marshy places.

Parts Used : Roots. Rhizome.

Reproductive Cycle : April–August.

Folk Uses : Small pieces of its roots tied around the neck to cure Kwashiorkar disease in children. A garland made from the rhizome put around the neck for cold and cough, expelling worms and warding off evils spirit especially in children.

Active Constituents : These are glucoside 'acorin', calamene tannin, mucilage, starch, vit C, fatty acids, sugar and calcium oxalate (roots), acoradin together with 2,4,5-trimethoxybenzaldehyde, 2,5-dimethoxybenzoquinone, glangin and sitosterol (rhizomes).

Trade : Sweet flag.

References : Bhattacharjee (2001), Chauhan (1999), Dutta and Nath (2000), Islam (1996), Khan and Khanum (2005), Rao *et al.* (1996), Sharma *et al.* (2003), Sinha and Sinha (2001), Sood *et al.* (2005).

***Actinopteris radiata* (Swartz) Link.**

Syn: *Acrostichum dichotomum* Forsk.; *Actinopteris australis* Linn. f.; *A. dichotoma* Bedd.

Family: Actinopteridaceae

Eng., H. and Sans. Names : *Eng.*: Peacock Tail.

H.: Marpath, Mayurshikha.

Sans.: Mayurashikhaa, Neelkantha, Shikhao, Varhishikhoa.

Distribution : Common in the Kumaon Hills and Nilgiri.

Part Used : Aerial Parts.

Reproductive Cycle : September–October.

Folk Use : Pieces of aerial parts of the plants in a sealed aluminium container smeared with turmeric powder worn around the waist for protection against effect of evil designs in Tamil Nadu.

Active Constituents : Active properties of aerial parts due to hentriacontane, hentriacontanol, β-sitosterol, its palmitate and its glucoside, an unidentified glucoside, glucose and fructose.

Trade : Sanjeevani.

References : Ambasta (1986), Chatterjee and Pakrashi (1997), Rastogi and Mehrotra (1970-1979), Sharma (2002, 2003), Sood *et al.* (2005).

***Adenanthera pavonina* Linn.**

Family: Fabaceae

Eng., H. and Sans. Names : *Eng.*: Coral wood tree, Red wood, sandal wood.

H.: Baragunci, Rakta Chandan, Rakta Kanchan.

Sans.: Kucandanah, Ranjaka, Tamrakah.

Distribution : Eastern sub-Himalayan tract. W. Ghats. Andamans.

Part Used : Seeds.

Reproductive Cycle : March–October.

Folk Use : Seeds strung and worn by women in many parts of India as ornaments and necklaces.

Active Constituents : Seeds contain dulcitol and stigmasterol.

References : Asolkar *et al.* (1992), Dhiman (2003), Balfour (1871), Mehra *et al.* (1975), Mittre (1997), Prajapati *et al.* (2006), Rastogi and Mehrotra (1970-1979).

***Adiantum caudatum* Linn.**

Syn: *A. incisum* Forsk.

Family: Adiantaceae.

Eng., H. and Sans. Names : *Eng.*: Tailed maiden hair.

H.: Marshikha.

Sans.: Mayurashikha, Nilkanthashikha.

Distribution : Throughout India.

Part Used : Roots.

Folk Use : A wreath of its roots tied to the left hand for facilitating easy delivery in expecting mothers; the wreath is removed as soon as delivery is over.

Active Constituents : Steroids, triterpenoids and flavonoids have been isolated from the plant.

References : Ambasta (1986), Chatterjee and Pakrashi (1997), Rastogi and Mehrotra (1980-1984), Singh and Narain (2008).

Aegiceras corniculatum L.

Syn: *Aechmanthera gossypina* (Nees) Nees; *A. tomentosa* Nees.; *A. majus* Gaertn .; *Rhizophora corniculata* Linn.; *Ruellia gossypina* Nees.

Family: Myrsinaceae

Eng. and H. Names : *Eng.*: Black mangrove, Goats horn mangrove.

H.: Halsi.

Distribution : Sunderbans. Andaman Islands.

Part Used : Flower.

Folk Use : In coastal region of Western India the pretty white flowers of this shrubby tree made into garlands.

References : Ambasta (1986), Anonymous (1948), Asolkar *et al.* (1992), Bennet (1986), Jain (1991), Mehra *et al.* (1975), Watt (1889).

Aegle marmelos (Linn.) Corr. *ex* Roxb.

Syn: *Belou marmelos* (Linn.) Lyons.; *Bilacus marmelos* (Linn.) Kuntze.; *Crateava marmelos* Linn.; *C. reliogosa* Ainslie.; *Feromia pellucida* Roth.

Family: Rutaceae

Eng., H. and Sans. Names : *Eng.*: Bael tree, Bengal Quince, Holy fruit, Indian Quince, Stone apple.

H.: Bael, Bel, Bilwa, Siriphal.

Sans.: Atimangaliya Bilva, Satyadharma, Shriphal, Tripatra.

Distribution : Submontane regions and plains almost throughout India.

Parts Used : Root. Bark. Wood. Thorns.

Reproductive Cycle : May–June.

Folk Uses : Roots collected in early morning on '*sankranti*' tied to the arm to prevent cholera. Tribals use the thorns of this tree for tattooing their bodies. Beads made from the bark and wood of this evergreen tree worn by low caste Hindus.

Active Constituents : Active principles of roots due to xanthotoxin, 6,7-di-o Me-Coumarin, scopoletin, tembamide, umbelliferone, marmesin, marmin, skimmianine and skimming.

Trade : Bel.

References : Anonymous (1948), Asolkar *et al.* (1992), Chauhan (1999), Jain (1994), Mehra *et al.* (1975), Rastogi and Mehrotra (1970-79), Sen and Behera (2007), Sood *et al.* (2005), Uniyal (2007), Watt (1889).

Aeschynomene aspera **Linn.**

Family: Fabaceae

Eng. and H. Names : *Eng.*: Sola pith.

H.: Sola.

Distribution : Wet places in Bengal and S. India.

Parts Used : Stem. Pith.

Folk Uses : Beads prepared from pith worn by aboriginal tribes as ear ornament or beautifully dyed in different shades for garlands for newly married couples and ornaments for idols. Soft stem also used for making wedding crowns for Hindus and veils for Muslims and for sola pith helmets.

References : Anonymous (1948), Dastur (1964), Francis (1983), Mehra *et al.* (1975), Watt (1889) Wealth (1948).

Aeschynomene indica **Linn.**

Syn: *A. diffusa* Willd.; *A. hispida* Willd.; *A. macropoda* DC.

Family: Fabaceae

Eng. and H. Names : *Eng.*: Hard sola.

H.: Laugauni, Shola.

Distribution : Throughout India upto 1,500m.

Part Used : Stem pith.

Folk Use : Stem pith used for making ear-tops. It is a cheaper substitute for *A. aspera.*

References : Ambasta (1986), Asolkar *et al.* (1992), Dastur (1964), Datta and Banerjee (1978), Jain (1991), Raju (2000).

Agaricus burkillii **(Massee.) Sacc. and Trodt.**

Family: Agaricaceae

Part Used : Fruit.

Folk Use : Sun dried fruiting body of fungus tied around the neck to cure goitre disease.

References : Rai *et al.* (1996), Sharma (2003).

Agave americana **Linn.**

Syn: *A. cantula* Roxb.

Family: Agavaceae

Eng., H. and Sans. Names : *Eng.*: American Aloe, Century Plant.

H.: Bans-keora, Bara kanwar, Kantala.

Sans.: Kalakantala.

Distribution : Cultivated in India.

Part Used : Fibre.

Reproductive Cycle : March–May.

Folk Use : Fibre used to string beads of *Aegle marmelos* wood.

References : Ambasta (1986), Sood and Thakur (2004), Watt (1889).

Ageratum conyzoides **Linn.**

Syn: *A. cordifoium* Roxb.

Family: Asteraceae

Eng., H. and Sans. Names : *Eng.*: Appa grass, Bastard agrimony, Conyzoid floss flowers, Goat weed, White weed.

H.: Koobhi, Visadodi.

Sans.: Visamustih.

Distribution : Along stream beds in the valleys in entire N.E. region, low moist situation and in abandoned jhum lands.

Part Used : Leaves.

Reproductive Cycle : Throughout the year.

Folk Use : Leaves placed over parturients belly to accelerate delivery, also bandaged over swollen parts of body to relieve pain.

Active Constituents : Leaves rich in stigmast 7-en-3-ol (major component), quercetin, kaempferol, fumaric and caffeic acids.

References : Asolkar *et al.* (1992), Jain and Sikarwar (1998), Khan and Khanum (2005), Kumar (2002), Prajapati *et al.* (2006), Rawat and Chowdhury (1998), Sikarwar (2002).

Ajuga macrosperma Wall. *ex* Benth.

Family: Lamiaceae

Distribution	:	Throughout Himalaya, Meghalaya, Upper Gangetic plains–Bihar and Tamil Naidu: 300–2,400m.
Part Used	:	Stem.
Folk Use	:	2-3 cm piece of its stem tied as an amulet to treat bodyache.
References	:	Girach (2001), Rastogi and Mehrotra (1980-1984).

Albizia lebbeck (Linn.) Willd.

Syn: *Acacia sirisa* Ham.; *A. speciosa* Willd.; *Albizia latifolia* Boivin.; *Mimosa lebbeck* Linn.; *M. sirisa* Roxb.; *M. speciosa* Jacq.

Family: Fabaceae

Eng., H. and Sans. Names	:	*Eng.*: East Indian walnut, Parrot tree, Siris tree, Sizzling tree. *H.*: Garso, Kalshish, Kalsir, Lasrin, Sirin, Siris, Tantia. *Sans.*: Bhandi, Kapitaka, Kapitana, Sirisha, Vishaghati, Vishahanta.
Distribution	:	Drier regions of India.
Parts Used	:	Bark. Flowers.
Reproductive Cycle	:	April–May.
Folk Use	:	Bark used for tattooing. Flowers used by N.Indian women to decorate their ears in summer. Women worn flowers as adornments.
Active Constituents	:	Tannin of condensed type-D-catechin, lebbecacidin (8,3′,4′-tetrahydroxyflavin-3,4-diol), isomers of leucocyanidin (5,7,3′,4′-tetrahydroxy flavan-3,4-diol) and friedelin are main colouring pigments of bark.
Trade	:	Indian walnut.
References	:	Jain (1995), Jain and Sharma (1996), Mehra *et al.* (1975), Sharma (2003), Sinha and Sinha (2001).

Aleurites moluccana (Linn.) Willd.

Syn: *A. ambinux* Pers.; *A. cordifolia* Steud.; *A. triloba* J.R. and G.Frost.; *Camirum cordifolium* Gaertn.; *Jatropha moluccana* Linn. .

Family: Euphorbiaceae

Eng., H. and Sans. Names	:	*Eng.*: Belagum walnut, Bengal walnut, Candlenut tree, Indian walnut. *H.*: Akola, Akrot, Jangliakrot.

Sans.: Akharota, Akhota, Akshota, Asphotaka, Kandarala, Karparala, Madanabhaphala, Parvatiya, Phalasneha, Pritakchhada, Rekhaphala, Vrittaphala.

Distribution : Various parts of India; found wild in S. India and Assam.

Part Used : Seed.

Reproductive Cycle : Flowers (March–April) and Fruits later.

Folk Use : Seed oil used for colouring eyelids.

Active Constituents : Seeds characterized by the occurrence of lipids and protein.

Trade : Jangli Akhrot.

References : Ambasta (1986), Chatterjee and Pakrashi (1997), Jain (1991), Kirtikar and Basu (1984), Prajapati *et al.* (2006).

Allium cepa* Linn.

Syn: *A. cumaria* Herb.

Family: Liliaceae

Eng., H. and Sans. Names : *Eng.*: Onion.

H.: Piyaz, Pyaj.

Sans.: Dirghapatra, Mahakanda, Nripapriya, Palandu, Rajapriya, Rochaka.

Distribution : Cultivated everywhere.

Part Used : Bulb.

Reproductive Cycle : Flowering in December–January.

Folk Uses : A small piece of bulb tied on body to cure boils. Tribals also keep bulb in the pocket during summers for protection against the effect of hot winds.

Active Constituents : Chief ingradients of bulbs are polyphenols, protocatechuic, caffeic, ferulic acids and propenyl sulfenic acid .

Trade : Common onion.

References : Ambasta (1986), Asolkar *et al.* (1992), Jadhav (2008), Rastogi and Mehrotra (1960-1969, 1980-1984), Sen and Behera (2007), Sharma (2003).

Allium sativum* Linn.

Family: Liliaceae

Eng., H. and Sans. Names : *Eng.*: Churls treacle, Garlic, Poor man's treacle.

H.: Lahsan, Lasun.

		Sans.: Arishtha, Bhutabhna, Grinjana, Lasuna, Rasona, Shuklakanda, Vatari.
Distribution	:	Cultivated as a kitchen garden crop in most parts of country.
Part Used	:	Cloves.
Reproductive Cycle	:	May–July.
Folk Use	:	A necklace comprising of 7-21 cloves worn for a week around neck with a sacred thread to cure typhoid, jaundice and whooping cough; it is believed that the size of necklace increases as the patient gets cured.
Active Constituents	:	Cloves comprise of allin, allicin, sativin, seordinine, scordine, anthocyanins, glycosides, querectin, vitamins like–thiamine, riboflavin and niacin.
Trade	:	Garlic.
References	:	Ambasta (1986), Chauhan (1999), Jadhav (2008), Sharma (2003), Singh and Pandey (1998), Sood *et al.* (2005), Watt (1889).

Aloe barbadensis Mill.*

Syn: *A. abyssinica* Lam.; *A. chinensis* Baker.; *A. indica* Royle.; *A. littoralis* Koening; *A. perfoliata* var. *vera* Linn.; *A. vera* (Linn.) Webb. and Berth., *A. vulgaris* Lam.

Family: Liliaceae

Eng., H. and Sans. Names	:	*Eng.*: Aloe, Barbados aloe, Common Indian aloe, Curcago aloe, Jaffarabad aloe.
		H.: Ghiguvara, Ghikumari, Gvarapatha, Kumari, Kuvarapatha.
		Sans.: Adala, Ajara, Amara, Aphala, Bhupatri, Brahmaghni.
Distribution	:	Kerala, Tamil Naidu. Hotter parts of India, Semi-arid states.
Part Used	:	Leaves.
Reproductive Cycle	:	July–November.
Folk Use	:	Bhils of Barmer district tie its leaves on affected portion to cure dropsy; also tied on waist to relieve pain of sciatica and backache.
Active Constituents	:	Isolated constituents from leaves are barbaloin aloesin, aloesone β-sitosterol, malic, citric and tartaric acids.
Trade	:	Indian aloe.

References : Ambasta (1986), Bhatt *et al.* (2003), Chauhan (1999), Khan and Khanum (2005), Prajapati *et al.* (2006), Rastogi and Mehrotra (1970-1979), Sharma (2003), Singh and Pandey (1998), Sood *et al.* (2009).

Alpinia allughas **(Retz.) Rosc.**

Syn: *A. nigra* (Gaertn.) Burtt.; *Zingiber nigrum* Gaertn.

Family: Zingiberaceae

Sans. Name : *Sans.*: Taraka.

Distribution : S. Konkan.

Parts Used : Flowers. Stem.

Reproductive Cycle : May–January.

Folk Use : Flowers used as adornments. Peel of its stem tied on the limbs for curing twingling sensation.

References : Bennet ((1986), Chakraborty (2003), Gogoi and Borthakur (1991).

Alstonia scholaris **(Linn.) R. Br.**

Syn: *Echites scholaris* Linn.

Family: Apocynaceae

Eng., H. and Sans. Names : *Eng.*: Devil's tree, Dita bark tree, Shaitan wood.

H.: Chatiun, Saitan ka jhad, Satiun, Satni, Satwin.

Sans.: Ayugmachchhada, Bahuparna, Saptaparnah, Vidha, Vinyaka.

Distribution : Found in sub–Himalayan tract, W. Bengal, Bihar, Peninsular India, Andamans.

Part Used : Twigs.

Reproductive Cycle : December–July.

Folk Use : Twigs worn as a ring during dance.

Trade : Chattim.

References : Ambasta (1986), Jain (1991), Khan and Khanum (2005), Prajapati *et al.* (2006), Rao (1997).

Alternanthera sessilis **(Linn.) DC.**

Syn: *A. denticulata* R. Br.; *A. nodiflora* R. Br.; *A. repens* Gmel.; *A. triandra* Lam.; *Gomphrena sessilis* Linn.

Family: Amaranthaceae

Eng., H. and Sans. Names : *Eng.*: Khaki weed, Sessile alternanthera.

H.: Gudrisaq, Lancheri.

Sans.: Giojihra, Matsyaksi, Patturah.

Distribution	:	Warmer parts of India.
Part Used	:	Root.
Reproductive Cycle	:	Flowering and Fruiting throughout the year.
Folk Use	:	Fresh root of the plant made to touch the cataract eyes five times a day for curing the disease; roots buried in water after every treatment.
Active Constituents	:	Roots rich in A and B spinasterols and lupeol.
References	:	Ambasta (1986), Bennet (1986), Khan and Khanum (2005), Prajapati *et al.* (2006), Rastogi and Mehrotra (1970-1979), Singh and Maheshwari (1992).

Alysicarpus vaginalis (Linn.) DC.

Syn: *A. diversifolius* Wall.; *A. rubibarna* Wall.; *A. vaginalis* var. *nummularifolius* (Willd.) Miq.; *Hedysarum vaginales* Linn.

Family: Fabaceae

Eng. Names	:	*Eng.*: Alyce clover.
Distribution	:	Throughout India.
Part Used	:	Root.
Reproductive Cycle	:	August–November.
Folk Use	:	Root pulled out in one breath and hung around the neck to stop hiccups in children.
References	:	Bennet (1986), Hembrom (1996), Jain *et al.* (1991).

Amanita muscaria (Fr.) S.F. Gray.

Syn: *Agaricus muscarius* Linn.; *A. nobilis* Bolton.

Family: Amanitaceae

Eng. Names	:	*Eng.*: Aga, Fly Agaric.
Distribution	:	Throughout the temperate regions of N. Hemisphere.
Reproductive Cycle	:	Fruiting occurs in summer.
Folk Use	:	Amulet of plant worn as adornment.
References	:	Wasson (1971).

Ammania pentandra Roxb.

Family: Lythraceae

Distribution	:	Moist places throughout tropical India.
Part Used	:	Whole Plant.
Folk Use	:	An amulet of the whole plant around neck cures fever in children below one year.
References	:	Hembrom (1996), Hooker (1897).

Ammannia peploides **Spreng.**

Family: Lythraceae

Distribution	:	Rice fields. Throughout the plains. Warm places in the lower mountains.
Part Used	:	Whole Plant.
Folk Use	:	Whole plant tied around the neck to cure fever in children below one year.
References	:	Hembrom (1996), Hooker (1897).

Ampelocissus latifolia **(Roxb.) Planch.**

Syn: *Vitis glabrata* Heyne.; *V. indica* Wall.; *V. latifolia* Roxb.; *V. montana* Lawson.; *V. zeylanica* Russell.

Family: Vitaceae

Eng., H. and Sans. Names	:	*Eng.*: Wild grape.
		H.: Jungli draksh, Pani bel.
		Sans.: Sitalata.
Distribution	:	Greater parts of India upto 1,200m in the Himalaya.
Part Used	:	Stem.
Reproductive Cycle	:	June–October.
Folk Use	:	A necklace of 7-21 pieces of its stem tied with a sacred thread around neck for a week to cure typhoid; it is believed that with the increase in size of necklace the patients get cured.
References	:	Ambasta (1986), Bennet (1986), Jadhav (2008), Khan and Khanum (2005).

Anadendrum paniculatum **DC.**

Syn: *Echites coriacea* Wall.; *E. manubriata* Wall.; *E. paniculata* Roxb.; *E. polyantha* Wall.

Family: Araceae

Distribution	:	Deccan Peninsula. W. Ghats from the Konkan southwards.
Part Used	:	Seeds.
Folk Use	:	Seeds made into necklaces on the Andaman Islands.
References	:	Radcliffe-Brown (1922), Sood *et al.* (2005).

Anisophyllea zeylanica R. Br.

Family: Rhizophoraceae

Distribution : Found in Andaman Islands.

Part Used : Seeds.

Folk Use : Seeds of several species strung on Great Andaman.

References : Ambasta (1986), Radcliffe-Brown (1922).

Annona squamosa Linn.

Syn: *Annona asiatica* Linn.

Family: Annonaceae

Eng., H. and Sans. Names : *Eng.*: Custard apple, Sugar apple, Sweet sop.

H.: At, Atasitaphal, Sharifa, Sitaphal.

Sans.: Agrimakhya, Gandagatra, Subha, Seethaphalam, Vaidehivallabha.

Distribution : Hilly tracts. Wastelands.

Part Used : Bark.

Reproductive Cycle : March–September.

Folk Use : Bark removed from a small twig in one breath tied to the hand of the crying baby for curing 'chinta'–a condition in which child cries unendingly and refuses the breast milk, within 5-7 minutes the child stops crying and become quite.

Active Constituents : Isolated ingredients of bark are corydine, isocorydine, anonaine, glaucine, boneal, camphor and a new monoterpene.

Trade : Sitaphal.

References : Ambasta (1986), Anonymous (1948), Asolkar *et al.* (1992), Chatterjee and Pakrashi (1997), Hembrom (1996), Rastogi and Mehrotra (1970-1979).

Anogeissus latifolia (Roxb. *ex* DC.) Wall. *ex* Guill. and Perr.

Syn: *A. latifolia* var. *glabra* Cl.; *A. latifolia* var. *tomentosa* Haines.; *A. latifolia* var. *villosa* Cl.; *Conocarpus latifolia* Roxb.

Family: Combretaceae

Eng., H. and Sans. Names : *Eng.*: Axle wood, Button tree, Dindiga tree, Ghatti tree.

H.: Bakla, Bakli, Bankli, Dhaoya, Dharta, Dhau, Dohu.

Sans.: Baka, Dhava, Dhurandhara, Gaura, Ghata, Kashaya, Pandura.

Distribution	:	Throughout the greater parts of India.
Part Used	:	Wood.
Reproductive Cycle	:	June–January.
Folk Use	:	Bhils of Dungarpur district tie an amulet comprising of a piece of its wood, a piece of leather and few hairs of a woman around the neck of cattle for warding off the effect of evil eyes.
Active Constituents	:	Heart wood contains gallic acid, ellagic acid, quercetin and myricetin.
Trade	:	Dhawa.
References	:	Ambasta ((1986), Anonymous (1948), Asolkar *et al.* (1992), Bennet (1986), Kirtikar and Basu (1984), Singh and Pandey (1998).

***Anthocephalus chinensis* (Lamk.) A. Rich *ex* Walp.**

Syn: *A. cadamba* (Roxb).; *A. indicus* Rich.; *Cephalanthus chinensis* Lamk.; *Neolamarckia cadamba* (Roxb.) Bosser.

Family: Rubiaceae

Eng., H. and Sans. Names	:	*Eng.*: Kadam, Wild cinchona. *H.*: Kadam, Kadamba. *Sans.*: Ashokari, Dasa, Deva, Haripriya, Kadamba, Kutsitangah, Priya.
Distribution	:	Sub-Himalayan tract from Sikkim eastwards to Bengal, E. Coast (Andhra Pradesh), W. Ghats.
Part Used	:	Flowers.
Reproductive Cycle	:	May–September.
Folk Use	:	In the rainy season, N. Indian women put the golden yellow flowers of this moderate–sized tree on their fore heads for adornment and as a symbol of their happy married life. Flowers worn as adornments by women.
Active Constituents	:	Flowers possess linalool, geraniol, geranyl acetate, linalyl acetate, alpha-selinene, 2-nonanol and beta phellandrene.
Trade	:	Kadam.
References	:	Bennet (1986), Chandra (1997), Dhiman (2003), Dymock (1889a), Khare (2004), Mehra *et al.* (1975), Sood *et al.* (2005), Prajapati *et al.* (2006).

***Antidesma diandrum* (Roxb.) Heyne *ex* Roth.**

Syn: *A. acidum* Retz.; *A. ghaesembilla* sensu Hk.f.

Family: Euphorbiaceae.

H. Names	:	*H.*: Amari, Amta, Dhakki, Katma, Khatua, Matta-ara.
Distribution	:	Himalaya: Kumaon eastwards–Sikkim. Peninsular India.
Part Used	:	Root.
Reproductive Cycle	:	May–October.
Folk Use	:	Roots tied to both legs and hands to cure polio in children; also its root pulled out in one breath hung around neck to cure toothache.
References	:	Ambasta (1986), Bennet (1986), Asolkar *et al.* (1996), Hembrom (1996), Rastogi and Mehrotra (1970-1979), Varghese (1996).

***Aporusa dioica* (Roxb.) Muell.**

Syn: *Alnus dioica* Roxb.; *A. octandra* (D. Don) Vickery; *A. roxburghii* Baill.; *A. villosula* Kurz.; *Myrica octandra* Buch. Ham. *ex* D. Don.

Family: Euphorbiaceae

Distribution	:	Tropical valleys of E. Himalaya.
Part Used	:	Bark.
Reproductive Cycle	:	February–May.
Folk Use	:	Used for tattooing.
References	:	Anonymous (1948), Bennet (1986), Asolkar *et al.* (1992), Chandra (1997), Jain and Borthakur (1980), Saklani and Jain (1994).

***Aquilaria agallocha* Roxb.**

Syn: *A. malaccensis* Roxb.; *A. ovata* Cav.

Family: Thymelaeaceae

Eng., H. and Sans. Names	:	*Eng.*: Aloe wood, Eagle-wood. *H.*: Agar. *Sans.*: Agaru, Krsnaguruh.
Distribution	:	Forests of Arunachal Pradesh, hilly tracts of Assam, Nagaland, Manipur and Tripura.
Part Used	:	Beads (Wood).
Reproductive Cycle	:	Flowering (May–June) and fruiting (July–August).

Folk Use	: Beads employed for necklaces and rosaries.
Trade	: Indian Eagle wood.
References	: Ambasta (1986), Dastur (1964), Sood *et al.* (2005), Watt (1889).

Areca catechu Linn.

Syn: *A. faufel* Gaertn.; *A. hortensis* Lour.

Family: Arecaceae

Eng., H. and Sans. Names	: *Eng.*: Areca nut palm, Areca palm, Betel nut, Catechu Palm, Pinang Palm. *H.*: Supari, Suppari, Supyri. *Sans.*: Akota, Chhataphala, Chikkana, Ghonta, Gubak, Kramuka, Puga-phalam, Tantusara.
Distribution	: S.W. and N.E. regions of the country (Kerala, Karnataka, Assam, W. Maharashtra, Tamil Nadu).
Part Used	: Nuts.
Reproductive Cycle	: Flowering in April–June.
Folk Uses	: Nuts of this Malayan slender–stemmed palm converted into fancy-shaped, polished beads for adornment as necklaces. Groom ties pierced seed on bride's wrist.
Active Constituents	: Nuts contain–(+) catechin, (-) epicatechin and procyanidin A-1, B-1 and B-2.
References	: Dastur (1964), Dymock (1889b), Mehra *et al.* (1975), Sood *et al.* (2009).

Argemone mexicana Linn.

Syn: *A. leiocarpa* Greene.

Family: Papaveraceae

Eng., H. and Sans. Names	: *Eng.*: Gamboge thistle, Jamaica yellow thistle, Mexican poppy, Prickly poppy. *H.*: Biladhutura, Brahmadundi, Kutaila, Satyanashi. *Sans.*: Hemadugdha, Kanchani, Katuparni, Pitapushpa, Suvarna, Svarnaksiri.
Distribution	: Throughout India.
Part Used	: Root.
Reproductive Cycle	: February–May.

Folk Use	:	Ring of the fresh root placed in hair of the expecting mother to facilitate delivery; ring removed immediately after delivery.
Active Constituents	:	Protopine, berberine, resin and a toxic principle are the chief constituents of roots.
Trade	:	Chok.
References	:	Ambasta (1986), Asolkar *et al.* (1992), Chatterjee and Pakrashi (1997), Khan and Khanum (2005), Saren *et al.* (2000), Sinha and Sinha (2001), Uniyal (2007).

Argyreia nervosa (Burm. f.) Boj.

Syn: *A. speciosa* (L.f.) Sweet.; *Convolvulus nervosus* Burm. f.; *C. speciosa* Linn.; *Ipomoea speciosa* (L.f.) Pers.; *Lettsomia nervosa* (Burm. f.) Roxb.; *Rivea nervosa* (Burm.f.) Hall. f.

Family: Convolvulaceae

Eng., H. and Sans. Names	:	*Eng.*: The elephant creeper, Wolly morning glory. *H.*: Samandar ka pat, Samandar phaind, Samandarsol. *Sans.*: Ajantri, Bastaantri, Chhagala, Dridhadaru, Dirgha, Murva, Samudrepalaka, Vriddhadara.
Distribution	:	All over N.E. region.
Part Used	:	Leaf.
Reproductive Cycle	:	August–September.
Folk Use	:	Leaves worn as a cap to cure headache; upper surface of the leaf bandaged over the boil for suppuration.
Active Constituents	:	Properties of leaves attributed to 1-triacontanol, β-sitosterol, epifriedelinol, quercetin, kaempferol and 3-O-rhamnoside.
References	:	Asolkar *et al.* (1992), Bennet (1986), Bhatt *et al.* (2003), Khan and Khanum (2005), Rastogi and Mehrotra (1990-1994), Sen and Behera (2007).

Arnebia benthamii (Wall. *ex* G.Don.) Johnston

Syn: *Macrotomia benthamii* A. DC.; *Echium benthami* Wall.; *Leptanthe macrostachya* Klotzsch

Family: Boraginaceae

H. Name	:	*H.*: Ratanjot.
Distribution	:	Found 3,300–4,350 m on the Pir–Panjal, W. Himalaya (Kumaon–Kashmir).
Part Used	:	Roots.

Reproductive Cycle : Flowering in July–August.

Folk Uses : Purple dye extracted by boiling the underground parts on low flame used for colouring hand, by Gaddies

Active Constituents : Naphathaquinones–shikonin and alkannin are main colouring pigments; their derivatives–deoxyshikonin, β-dimethylacrylshikonin, acetylshikonin, tetraacryl-shikonin, β-hydroxy–isovalerylshikonin, shikonofuran also responsible for reddish violet colour.

References : Ambasta (1986), Chauhan (1999), Jain (1991), Prajapati *et al.* (2006), Rastogi and Mehrotra (1998), Sood *et al.* (2009).

Arnebia guttata Bunge

Syn: *A. tibetiana* Kurz.

Family: Boraginaceae

Distribution : Kashmir (2,100–3,600m).

Part Used : Roots.

Folk Uses : Roots yield a red coloured fat–soluble edible dye for colouring lips.

Active Constituents : Naphthoquinones–shikonin and alkannin are main colouring pigments; their derivatives–deoxyshikonin, acetylshikonin, shikonin, β–hydroxyisovaleryl shikonin and tetraacryl shikonin also responsible for colour.

References : Ambasta (1986), Asolkar *et al.* (1992), Sood *et al.* (2009).

Artemisia nilagirica (Clarke.) Pamp.

Syn: *A. indica* Willd.; *A. vulgaris* auct.(non Linn.); *A. vulgaris* var. *nilagirica* C.B. Clarke.

Family: Asteraceae

Eng., H. and Sans. Names : *Eng.*: Fleabane, Indian wormweed, Motherwort, Mugwort.

H.: Dona, Gathivana, Majtari, Mastaru.

Sans.: Barha, Damanaka. Davanam, Naagdamani, Valamota.

Distribution : Hilly regions of India. Mount Abu in Rajasthan. W. Ghats, Konkan southwards to Kerala.

Parts Used : Aerial parts. Leaves. Plant.

Reproductive Cycle : August–November.

Folk Uses : Aerial parts tied to body for curing bodyache. Leaves used as bed sheet for relief from body pain. Dried plant kept under the pillow for warding off ghosts and evil spirits in house.

Active Constituents : These are trideca–1,3,5-triene-7,9,11-triyne, artemisia ketone (roots), a pentacyclic triterpene–fernenol (rhizomes), arabinose, glactose, glucose, maltose, raffinose, ribose, sucrose (stem), threonine, leucine, isoleucine, glycine, glutamic acid and arginine (leaves).

Trade : Nagdona.

References : Bennet (1986), Dhiman (2003), Khan and Khanum (2005), Rastogi and Mehrotra (1960-1969, 1980-1984), Rawat and Chowdhury (1998), Saklani and Jain (1994), Sood and Sudershna (2008), Sood *et al.* (2005).

Asparagus gonocladus **Baker.**

Family: Liliaceae

Eng., H. and Sans. Names : *Eng.*: Piri karijeri, Wild carrot.

H.: Shakakul, Shatavir.

Sans.: Satavari.

Distribution : Peninsular India from Konkan southwards in the Ghats. Plains of India.

Part Used : Root.

Reproductive Cycle : June–September.

Folk Use : Roots hung around the neck to cure fever in children below one year.

References : Chandra (1997), Hembrom (1996), Khan and Khanum (2005), Sood and Sudershna (2008).

Asparagus racemosus **Willd.***

Syn: *A. acerosus* Wall.; *A. sarmentosus* Wall.

Family: Liliaceae

Eng., H. and Sans. Names : *Eng.*: Indian asparagus, Wild asparagus.

H.: Bojhidan, Chatwal, Sadabori, Satawar, Satmuli, Shatavir.

Sans.: Abhiru, Aheru, Atirasa, Atmagupta, Bahusutaa, Bhiru, Divya, Indivari, Jata, Madhura, Mula, Shatamuli, Supatra, Vaishnavi, Vari, Vrishya.

Distribution	:	N. India.
Parts Used	:	Leaves. Roots. Twigs.
Reproductive Cycle	:	October–December.
Folk Uses	:	Leaves used as adornments. Twigs worn in caps during festivals in Madhya Pradesh. Root piece tied around the ears to relieve vertigo due to excessive heat, headache and swelling in body; also kept under the pillow for induction of sleep in children upto one year. A necklace of root hung around neck to remove worms in the cattle wounds. And for curing typhoid, 7-21 pieces of its roots worn as a necklace for a week.
Active Constituents	:	Leaves rich in quercetin-3-glucuronide whereas roots contain Satavarin, a glucoside of sarsasapogenin, shatavarin-I, shatavarin-II and shatavarin-IV.
Trade	:	Satawar.
References	:	Ambasta (1986), Banerjee (1974), Gogoi and Borthakur (1991), Hembrom (1996), Khan and Khanum (2005), Kirtikar and Basu (1984), Rastogi and Mehrotra (1960-1969), Sen and Behera (2007), Sharma (2003), Sinha and Sinha (2001), Sood and Sudershna (2008).

Asplenium aethiopicum **(Burm. f.) Bech.**

Family: Aspleniaceae

Distribution	:	Throughout India especially in mountainous region.
Part Used	:	Aerial parts.
Folk Use	:	An amulet of aluminium foil smeared with turmeric powder and containing pieces of aerial roots worn around waist for protection against the effect of evil designs.
Reference	:	Sood *et al.* (2005).

Atylosia volubilis **(Blanco.) Gamble.**

Syn: *A. crassa* Prain.; *A. mollis* Baker (FBI) in part non. Benth.; *Cytisus volubilis* Blanco.

Family: Fabaceae

Distribution	:	Almost throughout India upto 900m.
Part Used	:	Stem.
Folk Use	:	Stem tied as an amulet on the waist to cure body swelling.
References	:	Bennet (1986), Girach and Aminuddin (1995), Jain *et al.* (1991).

Azadirachta indica A. Juss.

Syn: *Melia azadirachta* Linn.; *Melia indica* Brandis.

Family: Meliaceae

Eng., H. and Sans. Names	:	*Eng.*: Indian liliac, Margosa tree, Neem tree. *H.*: Neem, Nim, Nimb. *Sans.*: Arishta, Kakaphala, Nimba., Subhadra, Vishirnaparna.
Distribution	:	Occurring naturally in Deccan Peninsula; cultivated all over India near habitations and roadsides.
Parts Used	:	Leaves. Twigs. Beads.
Reproductive Cycle	:	March–August.
Folk Uses	:	Beads from stones used as an ornament for making necklaces and rosaries. Meenas and Damors place the leaves on the bed of the expecting lady with the belief to keep away the evil spirits. Leaves also spread over the bed of a child suffering from viral infections like chicken pox and cow pox. For relief against rheumatic pains and swelling, the patient advised to sit under the plant to take rest for 2 or 3 days and nights or sharp blows given with twigs of the plant. Also used for tattooing.
Active Constituents	:	Isolated ingridients are flavonone, meliacinsolannolide, solannin, azadirachtannin, tetranosterpenoid isoazaditolide, nimbocinolide, quercitin and β-sitosterol (leaves), meliacin-4-epinimbin, 7-deacety-17 β hydroxyl azadiradione, 2 new tetranortriterpenoids-nimbanal and salannol-3-acetate (seeds).
Trade	:	Neem tree.
References	:	Bhatt *et al.* (2003), Jain (1991, 1994, 1995), Jain and Sharma (1996), Kaplesh *et al.* (2004), Khan and Khanum (2005), Kirtikar and Basu (1984), Mehra *et al.* (1975), Nargas and Trivedi (2000), Punjani (2002), Rastogi and Mehrotra (1980-1984), Sharma (2003), Singh and Pandey (1998), Sood *et al.* (2005).

Baccaurea courtallensis Muell.

Family: Euphorbiaceae.

Distribution	:	W. Ghats from N. Kanara southwards.
Part Used	:	Fruits.
Folk Use	:	Fruits used for making adornments.
References	:	Gogoi and Borthakur (1991), Rastogi and Mehrotra (1970-1979).

Baccaurea sapida **(Roxb.) Muell. Arg.**

Syn: *B. macrostachya* (W. and A.) Hk.f.; *B. ramiflora* Lour.; *B. wrayi* King *ex* Hk.f.; *Pierardia sapida* Roxb.

Family: Euphorbiaceae

H. Names	:	*H.*: Lutco, Kataphal.
Distribution	:	Wild or cultivated in sub-Himalayan regions; Andamans, chiefly in moist tropical forests.
Part Used	:	Bark.
Reproductive Cycle	:	March–June.
Folk Use	:	Bark used for tattooing.
Active Constituents	:	Bark contains friedelin, epifriedelinol, Me-betulinate, β-sitosterol and an unknown ester.
References	:	Ambasta (1986), Bennet (1986), Asolkar *et al.* (1992), Jain and Borthakur (1980), Rastogi and Mehrotra (1960-1969), Saklani and Jain (1994).

Bambusa vulgaris **Schrad. *ex* Wendl.**

Family: Poaceae

Eng., H. and Sans. Names	:	*Eng.*: The golden bamboo, The pantropical tiger bamboo, The yellow and green striped bamboo. *H.*: Basinibans. *Sans.*: Bahupallava.
Distribution	:	Wild in the warmer parts of India.
Part Used	:	Culms.
Folk Uses	:	Used as ornaments of arm and ear. Rings prepared from the split culms of this species put into ear perforations by the Tunkul–Naga tribes of Manipur and beads used by the Andaman Islands. Maria gonds and Hos make bangles and wedding crowns in the north.
References	:	Ambasta (1986), Anonymous (1948), Hooker (1897), Jain (1991), Majumdar (1950), Mehra *et al.* (1975), Radcliffe-Brown(1922), Reddy (1956).

Baphicacanthus cusia **(Nees.) Bremek.**

Syn: *Strobilanthes cusia* Imlay.; *S. flaccidifolius* Nees.

Family: Acanthaceae

Eng. Name	:	*Eng.*: Assam indigo.
Distribution	:	N.E. India.

Part Used : Leaf.

Folk Use : A blue dye from the juice of leaf used for making tattoos on face by Mikir women.

Active Constituents : Leaves characterized by the occurrence of quingdainone along with tryptanthrin.

References : Asolkar *et al.* (1992), Borthakur (1997), Jain (1991), Rastogi and Mehrotra (1980-1984), Saklani and Jain (1994).

Barleria prionitis **Linn.**

Syn: *B. hystrix* Linn.; *B. pubiflora* Benth.; *Prionitis hystrix* Miq.; *P. pubiflora* Miq.

Family: Acanthaceae

Eng., H. and Sans. Names : *Eng.*: Thorn nail dye.

H.: Jhinti, Katsareya.

Sans.: Karunta, Karuntaka, Vajradanti.

Distribution : Himalaya–Sri Lanka.

Part Used : Flowers.

Reproductive Cycle : October–March.

Folk Use : Flowers used for hair adornments.

Active Constituents : Flowers contain scutellarein-7-rhamnoglucoside.

References : Ambasta (1986), Dymock (1889a), Kumar and Nagiyan (2006), Sood *et al.* (2005).

Basella alba **Linn. var.** ***rubra*** **(Linn.) Stewart.**

Syn: *B. cordifolia* Lamk.; *B. rubra* Stewart.

Family: Chenopodiaceae

Eng., H. and Sans. Names : *Eng.*: Indian spinach, Malabar nightshade.

H.: Bon-poi, Lalbachlu, Myal-ki-bhaji, Poi, Sufed bachla.

Sans.: Kalambi, Madhushaka, Mohini, Potaki, Putika, Upoti, Valipodaki, Vishala, Vishvatulasi, Vrishchika-priya.

Distribution : Cultivated throughout N.E. states.

Part Used : Flowers.

Reproductive Cycle : March–December.

Folk Use : Flowers used for decorating hands.

References : Bennet (1986), Khan and Khanum (2005), Kirtikar and Basu (1984), Saklani and Jain (1994), Sood and Sudershna (2008).

Bauhinia racemosa **Lamk.**

Syn: *B. parviflora* Valh.; *B. spicata* Koenig.

Family: Fabaceae

Eng., H. and Sans. Names	:	*Eng.*: Camel's foot climber. *H.*: Ashta, Jallur, Kachnal, Maljhan, Malu. *Sans.*: Asmantaka, Malanjhana, Murva, Swetakanchan.
Distribution	:	Kashmir–Sikkim upto 1,500m; Gujarat southwards to Maharashtra and northern A.P.
Part Used	:	Stem bark, Tendrils. Flowers.
Reproductive Cycle	:	April–June.
Folk Uses	:	Necklace prepared from stem bark worn for a week to cure conjunctivitis. Dried tendrils worn as finger rings and as charm against dropsy by Santhal Tribes. Flower garlands worn by Chola kings.
Active Constituents	:	Principle constituents of bark are tannin, octacosane, β-amyrin and β-sitosterol.
References	:	Ambasta (1986), Asolkar *et al.* (1992), Chandra (1973), Jadhav (2008), Khan and Khanum (2005), Mehra *et al.* (1975), Sood *et al.* (2009).

Bauhinia vahlii **Wight. and Arn.**

Syn: *Phanera vahlii* Benth.

Family: Fabaceae

H. and Sans. Names	:	*H.*: Jalluar, Jallur, Mahi, Malghan, Malijan, Maljan. *Sans.*: Jallur, Maljhan.
Distribution	:	Assam, Bihar, M.P., Panjab.
Parts Used	:	Tendrils. Stem.
Reproductive Cycle	:	Flowering (March–April) and fruiting in winter.
Folk Uses	:	Dried tendril rings worn on fingers by Santhals. Stems used for loin–cloth belts by male Reddis.
References	:	Ambasta (1986), Furer-Haimendorf (1945), Sood *et al.* (2009), Watt (1889).

Begonia picta **Sm.**

Syn: *B. echinata* Royle.

Family: Begoniaceae

Distribution	:	Almost throughout India upto 2,000m.
Part Used	:	Leaf.

Reproductive Cycle : August–November.

Folk Use : Leaves used to impart red colour on palm during festivals in Manipur.

References : Asolkar *et al.* (1992), Chowdhery and Wadhwa (1984), Saklani and Jain (1994), Sood and Thakur (2004).

***Berberis wallichiana* DC.**

Syn: *B. asiatica* Griff.

Family: Berberidaceae

Eng. Names : *Eng.*: Wallich's barbery, Wallich's berberis.

Distribution : Temperate Himalaya, 2,700-3,350m.

Part Used : Spine.

Reproductive Cycle : March–November.

Folk Use : Spines used for tattooing by Apatani of Ziro.

References : Rawat and Chowdhury (1998), Saklani and Jain (1994).

***Bergenia ciliata* (Haw.) Sternb.**

Syn: *B. ligulata* (Wall.) Engl.; *Megasea ciliata* Haw.; *Saxifraga ciliata* Wall.; *S. ligulata* Wall.

Family: Saxifragaceae

Eng., H. and Sans. Names : *Eng.*: Rock foil.

H.: Pashanbhed.

Sans.: Ashmabhedaka, Bhimayojini, Nagabhita, Paashaanbheda, Shveta.

Distribution : Throughout temperate Himalaya (Kashmir–Bhutan). Khasi Hills.

Part Used : Leaves.

Reproductive Cycle : March–July.

Folk Use : Leaves kept on head under the cap for relieving fever.

Active Constituents : Leaves rich in flavonoids.

Trade : Pashanbheda.

References : Ambasta (1986), Bennet (1986), Chauhan (1999), Khare (2004), Sharma (2003), Sood and Sudershna (2008)

Biophytum apodiscias **(Turez.) Edgew. and Hk. f.**

Syn: *B. petersianam* Klotz.; *Oxalis apodiscias* Turez.

Family: Oxalidaceae

Distribution	:	W. Himalaya.
Part Used	:	Whole Plant.
Folk Use	:	Plant kept under pillow for induction of sleep in children.
References	:	Ambasta (1986), Bennet (1986), Hembrom (1996).

Biophytum reinwardtii **Walp.**

Syn: *B. reinwardtii* var. *metziana* Edge. and Hk. f.

Family: Oxalidaceae

Distribution	:	Kumaun–Arunachal Pradesh upto 1,800m and extending upto Peninsular India.
Part Used	:	Whole Plant.
Folk Use	:	Plant kept under pillow of children for induction of sleep.
References	:	Bennet (1986), Asolkar *et al.* (1992), Hembrom (1996).

Bixa orellana **Linn.**

Family: Bixaceae

Eng., H. and Sans. Names	:	*Eng.*: Annatto, Arnatto, Chinese dye tree, Monkey turmeric, Roucou. *H.*: Latkan, Sindhuriyan, Vatkana. *Sans.*: Karachchada, Raktabija, Raktapushpa, Sinduri, Shona pushpi.
Distribution	:	Cultivated throughout India.
Part Used	:	Seeds.
Reproductive Cycle	:	Flower (summer) and Fruit (winter).
Folk Use	:	Tribal girls used to redden their cheeks with seeds.
Active Constituents	:	Main ingradients of seeds are bixin, norbixin, β-carotene, cryptoxanthin, lutein, zeaxanthin and methylbixin.
Trade	:	Annatto.
References	:	Asolkar *et al.* (1992), Kirtikar and Basu (1984), Lalramnghinglova (1996), Rastogi and Mehrotra (1990-1994), Sood and Sudershna (2008).

Blastania garcinii **(Burm.f.) Cogn.**

Syn: *Ctenolepis garcini* C.B. Cl.

Family: Cucurbitaceae

Part Used	:	Seeds.
Reproductive Cycle	:	September–December.
Folk Use	:	An amulet of its seeds tied around the arm of children for protecting them from evil spirits by 'Saharia' tribals of Rajasthan.
References	:	Singh and Pandey (1998), Sood *et al.* (2005).

Boerhaavia diffusa **Linn.**

Syn: *B. erecta* Linn.; *B. repens* Linn.; *B. repens* var. *diffusa*. (Linn.) Hk. f.; *B. repens* var. *procumbens* (Roxb.) Hk. f.

Family: Nyctaginaceae

Eng., H. and Sans. Names	:	*Eng.*: Pigweed, Spreading hogweed. *H.*: Beshakapore, Biskhafra, Punura, Santhi, Thikri; *Sans.*: Bhauma, Kathillaka, Krishnakhya, Krura, Lohita, Narva, Nila, Punarnava, Raktakanda, Raktapunarvana, Sarini, Shyma, Sophaghni, Varshabhava, Vishaghni.
Distribution	:	Throughout India.
Part Used	:	Root.
Reproductive Cycle	:	During winter.
Folk Use	:	A garland of 7-21 pieces of its root worn around the neck for a week to cure typhoid and jaundice.
Active Constituents	:	Roots found to contain alkaloid punarnavine, C-methyl flavone, rotenoid analogues–boeravinone A-F, aminoacids, fatty acids, β-sitosterol, hentriacontanes.
Trade	:	Pigweed.
References	:	Bennet (186), Jadhav (2008), Jain *et al.* (1991), Khan and Khanum (2005), Kirtikar and Basu (1984), Ranjan (1996), Sharma (2003), Rastogi and Mehrotra (1970-1979, 1980-1984), Sinha and Sinha (2001).

Bombax ceiba **Linn.**

Syn: *B. hexaphyllum* Roxb.; *B. malabaricum* DC.; *Gossampinus malabarica* (DC.) Merill.; *Salmalia malabarica* DC.

Family: Bombacaceae

Eng., H. and Sans. Names	:	*Eng.*: Cotton tree, Red silk cotton, Silk cotton tree. *H.*: Kantisembal, Pagun, Semal, Shimbal. *Sans.*: Bahuvirya, Nirgandhapuspi.

Distribution	:	Hotter parts all over India.
Part Used	:	Bark.
Reproductive Cycle	:	January–May.
Folk Use	:	Bark used for tattooing; its garland tied around the neck of cattle for wiping away the maggots from the wounds of mouth.
Active Constituents	:	Constituents isolated from stem bark are lupeol, β-sitosterol, naphthoquinone and potassium nitrate.
Trade	:	Mochavasa, Supari-ka-phool.
References	:	Asolkar *et al.* (1992), Jain (1995), Jain and Sharma (1996), Khan and Khanum (2005), Kshirsagar and Singh (2007), Rastogi and Mehrotra (1970-1979), Sinha and Sinha (2001), Sood and Sudershna (2008), Sood *et al.* (2009).

Borassus flabellifer **Linn.**

Syn: *B. aethiopum* Mart.; *B. dichotomus* White; *B. flabelliformis* Roxb.; *Lontanus domestica* Rumph.

Family: Arecaceae

Eng., H. and Sans. Names	:	*Eng.*: Brab tree, Char palm, Desert palm, Fan palm, Palmyrah palm. *H.*: Tal, Tala, Tar, Tari, Tarkajhar. *Sans.*: Asavardu, Chirayu, Dirghadru, Drumeshvara, Patri, Talah, Tamsi, Taruraja.
Distribution	:	Coastal areas of Peninsular India, Bihar and Bengal.
Part Used	:	Leaves.
Reproductive Cycle	:	March–June.
Folk Use	:	Santals cut the leaves into bracelets for adornments.
Trade	:	Taal.
References	:	Ambasta (1986), Anonymous (1948), Gogoi and Borthakur (1991), Khan and Khanum (2005), Kirtikar and Basu (1984), Sood and Sudershna (2008), Watt (1889).

Borassus ramiflora **Lour.**

Family: Arecaceae

Part Used	:	Fruits.
Folk Use	:	Fruits used as adornments.
Reference	:	Gogoi and Borthakur (1991).

Borreria articularis **L.f.**

Syn: *B. hispida* K. Schum.; *Spermacoce articularis* L.f.; *S. hispida* Linn.

Family: Rubiaceae

Eng., H. and Sans. Names	:	*Eng.*: Shaggy button weed. *H.*: Guthari, Madanaghanti. *Sans.*: Madanaghanta, Madanaghanti.
Distribution	:	Himalaya from the neighbourhood of Shimla.
Part Used	:	Root.
Reproductive Cycle	:	July–December.
Folk Use	:	Root pulled out in one breath and hung around neck to cure toothache.
Active Constituents	:	β sitosterol, ursolic acid and d-mannitol.
References	:	Ambasta (1986), Bennet (1986), Hembrom (1996).

Boswellia serrata **Roxb. *ex* Colebr.**

Syn: *B. floribunda* Linn.; *B. serrata* var. *glabra* Roxb.; *B. thurifera* Colebr.; *Labinus thurifera* Colebr.

Family: Burseraceae

Eng., H. and Sans. Names	:	*Eng.*: Indian frankincense tree, Indian oilbanum tree. *H.*: Kundur, Loban, Salga, Salai, Sali, Selgond. *Sans.*: Kapitthaparni, Sallakee, Shalaki.
Distribution	:	Bihar, Deccan Peninsula, M.P., C. India, Eastern states.
Part Used	:	Bark.
Reproductive Cycle	:	Flowers (March–April) and Fruits (Winter).
Folk Use	:	Used for tattooing.
Active Constituents	:	Bark rich in carbohydrates, glucosides and β-sitosterol.
Trade	:	Salai.
References	:	Ambasta (1986), Anonymous (1948), Bennet (1986), Dhiman (2003), Jain (1995), Jain and Sharma (1996), Rastogi and Mehrotra (1960-1969).

Bougainvillaea spectabilis **Willd.**

Family: Nyctaginaceae

Eng. and H. Names	:	*Eng.*: Garden glory. *H.*: Bogenvillea.
Distribution	:	Cultivated in Indian gardens.

Part Used : Flowers.

Folk Use : Flowers made into garlands in Bombay.

Reference : Francis (1984).

Brassica alba **H.f. and T.**

Syn: *B. foliosa* Willd.; *Sinapsis alba* Linn.

Family: Brassicaceae

Eng., H. and Sans. Names : *Eng.*: White mustard.

H.: Safed rai.

Sans.: Shvetasarshapa, Siddhartha.

Distribution : Cultivated in most parts of India.

Part Used : Seeds.

Folk Use : Seeds kept under pillow to ward off evil spirits.

Active Constituents : Active ingridients of seeds are benzyl isothiocyanate, sinapine thiocyanate, β-sitosterol, sinapic acid, benzoic acid.

References : Anonymous (1948), Asolkar *et al.* (1992), Rastogi and Mehrotra (1985-1989), Sood *et al.* (2005).

Bryophyllum pinnatum **(Lam.) Kurz***

Syn: *B. calycinum* Salisb.

Family: Crassulaceae

Eng. and H. Names : *Eng.*: Air plant, Life plant, Miracle leaf.

H.: Zakham-haiyat.

Distribution : Temperate regions of Asia.

Part Used : Leaves.

Reproductive Cycle : December–March.

Folk Use : Leaves tied to cure boils and swelling on affected part.

Active Constituents : These are ferulic, caffeic and p-hydroxybenzoic acids, quercetin and kaempferol

References : Ambasta (1984), Rastogi and Mehrotra (1970–1979).

Bulbophyllum neilgherrense **Wight.**

Syn: *B. kaitense* Rchb.f.; *Cirrhopetalum nilgherrense* W.

Family: Orchidaceae

Distribution : Found in S. India mainly in W. Ghats.

Part Used : Aerial parts.

Folk Use	:	Aerial parts contained in an amulet of aluminium tied around waist by a cotton thread for warding off the effects of evil designs.
References	:	Bennet (1986), Sood *et al.* (2005).

Butea monosperma (Lamk.) Taub.

Syn: *B. frondosa* Koen. *ex* Roxb.; *Erythrina monosperma* Lamk.

Family: Fabaceae

Eng., H. and Sans. Names	:	*Eng.*: Bastard teak, Bengal kino, Butea gum, Flame of the forest. *H.*: Chichra, Dhak, Kankrei, Palas, Palash, Tesu. *Sans.*: Bijasneha, Ksharashreshtha, Palasha, Vatapotha.
Distribution	:	All over India, chiefly in the mixed or deciduous forests of Central and W. India.
Parts Used	:	Flowers. Root. Bark.
Reproductive Cycle	:	April–July.
Folk Uses	:	Flowers used as adornments in ear by Hindu women. Roots collected on Saturday and tied to the arm of women to get rid of barrenness. A large piece of bark bandaged with rope to heal the fractured leg of cattle.
Active Constituents	:	These are β-sitosterol, leucoantho-cyanide, amyrin, betulinic acid, stigma-sterol, palasonin (bark, root); 7 flavonoid glucosides, butrin, isobutrin, 3 glucosides–coreopsin, isocoreopsin and sulphurein (flowers).
Trade	:	Plashapapra, Dhaak phool, Bengal kino.
References	:	Bennet (1986), Chauhan (1999), Dhiman (2003), Gogoi and Borthakur (1991), Jain (1991, 1994), Jain *et al.* (1991), Mehra *et al.* (1975), Ramdas *et al.* (2000), Sen and Behera (2007), Sood *et al.* (2009), Upadhye and Kumbhojkar (2003).

Caesalpinia bonduc (Linn.) Roxb.

Syn: *C. bonducella* (Linn.) Fleming.; *C. crista* Linn.; *Guilandina bonduc* Linn.; *G. glabra* Mill.

Family: Fabaceae

Eng., H. and Sans. Names	:	*Eng.*: Bonduc nut, Fever nut, Indian filbert, Niekar, The mysore thorn. *H.*: Kanja, Karanju, Kanta-karanj. *Sans.*: Kantakini, Karanj, Kuberaksh, Putikaranja.

Distribution	:	Throughout the hotter parts of India particularly along the sea coast and upto 830m.
Part Used	:	Seeds.
Reproductive Cycle	:	April–July.
Folk Use	:	Bracelets and necklaces of seeds with silken red thread worn by the expected mother to prevent abortion and against witchcraft.
Active Constituents	:	Chief constituents of seeds are α caesalpin, β caesalpin, γ caesalpin, aspartic acid, arginine, citrulline, a new diterpine and bonducellin.
Trade	:	Katkaranja, Natkaranja, Saagargola.
References	:	Bennet (1986), Dhiman (2003), Jain (1991), Mehra *et al.* (1975), Sood *et al.* (2005, 2009), Watt (1889).

Cajanus cajan **(Linn.) Millsp.**

Syn: *C. indicus* Spreng.

Family: Fabaceae

Eng., H. and Sans. Names	:	*Eng.*: Congo pea, Pigeon pea, Red gram. *H.*: Arhar, Tur, Tuver. *Sans.*: Adhaki, Tuvari, Tuvarika.
Distribution	:	Vastly cultivated in India.
Part Used	:	Wood.
Reproductive Cycle	:	August–January.
Folk Use	:	Wood of this shrub cut into small beads for rosaries.
References	:	Ambasta (1986), Kirtikar and Basu (1984), Mehra *et al.* (1975), Varghese (1996), Watt (1889).

Caladium bicolor **(Ait.) Vent.**

Family: Araceae

Distribution	:	Cultivated for ornamental foliage throughout India.
Parts Used	:	Leaves. Shoots.
Reproductive Cycle	:	May.
Folk Use	:	Used as adornments.
Reference	:	Gogoi and Borthakur (1991).

Calamus erectus **Roxb.**

Family: Arecaceae

Distribution	:	Sikkim, Assam, Meghalaya, Manipur, W. Bengal.

Part Used : Stem.

Folk Use : Stem used for making ear and nose ornaments by the people of Arunachal Pradesh.

Reference : Saklani and Jain (1994).

Calamus floribundus **Griff.**

Family: Arecaceae

Distribution : N. E India.

Part Used : Stem.

Reproductive Cycle : March–October.

Folk Use : Stem employed for making ear and nose ornaments by Arunachal people.

References : Chandra (1997), Saklani and Jain (1994).

Calamus inermis **T. Anders.**

Family: Arecaceae

Distribution : N. E. India.

Part Used : Stem.

Folk Use : Rings of stem worn in legs by both men and women of N. E. India as ornaments.

References : Thomas *et al.* (1998).

Calamus latifolia **Roxb.**

Family: Arecaceae

Distribution : E. Bengal; Assam.

Part Used : Stem.

Folk Use : People use rings of stem in their legs as ornaments.

References : Anonymous (1950), Thomas *et al.* (1998).

Callicarpa arborea **Cl.**

Syn: *C. farinosa* Roxb.; *C. teconaefolia* Wall.; *C. villosa* Roxb.

Family: Verbenaceae

Eng., H. and Sans. Names : *Eng.*: Japanese beauty berry, Japanese callicarpa.

H.: Khoja.

Sans.: Bagodi.

Distribution : Upper Gangetic Plains; lower hills of Kumaon–Sikkim; Assam; Khasi Hills; Bengal.

Parts Used	:	Bark. Leaves.
Reproductive Cycle	:	April–August.
Folk Use	:	Bark and tender leaves chewed to impart pink colour to lips in Manipur.
Active Constituents	:	Bark rich in Me-betulinate, baurenol and β-sitosterol. Leaves contain-β-sitosterol, maslinic, oleanolic and ursolic acids.
References	:	Asolkar *et al.* (1992), Rastogi and Mehrotra (1970-1979), Saklani and Jain (1994).

Callicarpa macrophylla Vahl.

Syn: *C. incana* Roxb.; *C. roxburghii* Wall.

Family: Verbenaceae

H. and Sans. Names	:	*H.*: Daya.
		Sans.: Phalini, Vanita.
Distribution	:	N. E. India, ascending to 2,000m in Himalaya
Parts Used	:	Leaves. Root.
Reproductive Cycle	:	June–September.
Folk Use	:	Used for imparting pink colour to lips.
Active Constituents	:	Glycosides, phenolic compounds, flavonoids–luteolin, apigenin and its 7-glucuronides are main colouring components.
References	:	Asolkar *et al.* (1992), Rastogi and Mehrotra (1995), Sood and Thakur (2004), Sood *et al.* (2009), Watt (1972).

Calophyllum inophyllum Linn.

Syn: *C. bintagor* Roxb.

Family: Clusiaceae

Eng., H. and Sans. Names	:	*Eng.*: Alexadrian laurel, Indian laurel.
		H.: Sultan campa, Surpan, Surpunika.
		Sans.: Nagachama, Nameru, Punnagah, Surangi.
Distribution	:	On the sea shores of India, in Maharashtra, Karnataka, Orissa, Andamans; also cultivated in Indian plains.
Part Used	:	Flowers.
Reproductive Cycle	:	July–October.
Folk Use	:	Flowers used for making garlands.

Active Constituents : Flowers contain myricetin and quercetin from androecium and leucocyanidin from petals.

References : Ambasta (1986), Anonymous (1950), Chakrabarty and Balakrishnan (2003), Khan and Khanum (2005), Rastogi and Mehrotra (1960-1969, 1970-1979).

Calotropis gigantea **(Linn.) R. Br. *ex* Ait.**

Syn: *Asclepias gigantea* Linn.

Family: Asclepiadaceae

Eng., H. and Sans. Names : *Eng.*: Bowstring hemp, Gigantic milk weed, Gigantic swallow wort, Madar.

H.: Ak, Akada, Arka, Lalat, Madar, Mudar, Safed ak.

Sans.: Aditya, Arkah, Mandara, Ravi, Rupika.

Distribution : Common throughout India in warm and dry places.

Parts Used : Flowers. Leaves. Seeds. Latex.

Reproductive Cycle : Throughout the year.

Folk Uses : Flower garlands used for making ornaments. Flowers and leaves worn in hair by Maria Gonds. Pillow made of floss from seeds used for children suffering in earache. Latex used for colouring the skin.

Active Constituents : Active principles of flowers due to β-sitosterol, taraxasterol, α- and β-amyrins. Latex possesses gigantin, uscharidin, usoharin and enzymes calotropin D-I, callotropin D-II, protease, calotropain and β-calotropeol, gigantin, β–sitosterol and uscharidin uscharin [1].

Trade : Madar.

References : Banerjee (2000), Chauhan (1999), Jain (1991), Khan and Khanum (2005), Mehra *et al.* (1975), Nagar (1980), Sharma (2003), Sood *et al.* (2009), Watt (1889).

Calotropis procera **(Ait.) R.Br.***

Syn: *Asclepias procera* Willd.; *C. hamiltonii* W.; *C. heterophylla* Wall.; *C. procera* Willd.; *C. wallichii* Wight.

Family: Asclepiadaceae

Eng., H. and Sans. Names : *Eng.*: Dead sea apple, Milk weed, Swallow wort.

H.: Ak, Aakada, Madar, Safed ak.

Sans.: Alarka, Arka, Bimbora, Mandara, Shamblau, Vasuka.

Distribution	:	Warm dry places throughout India from Panjab to Western, Central and Southern India.
Parts Used	:	Leaves. Root. Bark. Floss of fruits.
Reproductive Cycle	:	February–May.
Folk Uses	:	Mature yellow coloured leaves bandaged on head in the form of cap and also kept at sole of foot to check the sunstroke. Leaves also tied over the affected part to cure mumps, body swelling and pain. Fresh bark bandaged on udder to cure cracking of teats in animals. Root collected in the early morning of sunday from north of the plant, and tied with a red thread on the affected part to relieve filaria. To relieve rheumatic pain the patient rest on pillows and mattresses stuffed with floss of fruits of the plant.
Active Constituents	:	These are quercetin-3-rutinoside and β-amyrin (roots, leaves), benzoyllinedone and benzoylisolineoline (root bark).
References	:	Bennet (1986), Dhiman (2003), Khan and Khanum (2005), Kumar *et al.* (2004), Punjani (2002), Rastogi and Mehrotra (1970-1979), Sen and Batra (1997), Sood *et al.* (2009).

Canna orientalis **Rosc.**

Syn: *C. chinensis* Willd.; *C. indica* Rosc.; *C. nepalensis* Wall. *ex* Bouche.

Family: Cannaceae

Eng., H. and Sans. Names	:	*Eng.*: Indian bead, Indian reed, Indian shot. *H.*: Sabbajaya, Sarvajjaya, Kewara. *Sans.*: Devakili, Kamakshi, Sarvajaya, Shilarambha, Vanakadali.
Distribution	:	Widely planted in India.
Part Used	:	Seeds.
Reproductive Cycle	:	January–February.
Folk Use	:	Seeds of this species strung as beads along with seeds of other plants for making necklaces and other ornaments.
References	:	Anonymous (1950), Bennet (1986), Jain (1991), Mehra *et al.* (1975), Mittre (1997), Sood *et al.* (2009), Wealth (1950).

Canna speciosa Sims.

Syn: *C. indica* var. *speciosa* (Sims.) Baker.

Family: Cannaceae

Part Used	:	Flowers.
Folk Use	:	Flowers used as adornments.
References	:	Bennet (1986), Gogoi and Borthakur (1991).

Cannabis sativa Linn.

Syn: *C. indica* ssp. *indica* var. *indica* (Lamk.) Wehmer.; *C. indica* Lamk.; *C. sativa* ssp. *indica* var. *kafiristanica* (Vavilov.) Small and Cronquist.

Family: Cannabinaceae

Eng., H. and Sans. Names	:	*Eng.*: Cannabis, Indian hemp, Marihuana, Soft hemp, True hemp. *H.*: Bhang, Charas, Ganja, Gur, Kinnab, Siddhi. *Sans.*: Ajaya, Ananda, Bhanga, Bhringi, Chapala, Ganjika, Hanga, Harsini, Jaya, Matta, Matuli, Mohini, Nili, Shivpriya, Vijaya, Yogini.
Distribution	:	Temperate and tropical regions of India.
Parts Used	:	Leaves. Shoots. Fiber.
Reproductive Cycle	:	August–September.
Folk Uses	:	Used for adornments. Fiber prescribed for stringing beads.
Active Constituents	:	N-acetylglucosamine and N–acetylgalactosamine have been identified from leaves and shoots.
Trade	:	True Hemp.
References	:	Bennet (1986), Chauhan (1999), Dhiman (2003), Gogoi and Borthakur (1991), Rastogi and Mehrotra (1970-1979), Sharma (2003), Shende (1949), Sood and Sudershna (2008).

Capparis aphylla Roth.

Syn: *C. decidua* Edgew.

Family: Capparidaceae

Eng., H. and Sans. Names	:	*Eng.*: Caper plant. *H.*: Karel, Karer, Kurrel, Lete. *Sans.*: GudhapatraKarira, Krishashakha, Mriduphala, Suphala, Vishvakpatra.

Distribution	:	Distributed in Kutch, W. Rajasthan, Panjab, Upper Gangetic Plains, M.P., Gujarat, Deccan, Tinevelly.
Reproductive Cycle	:	March–April.
Folk Use	:	Juice used to prepare etched carnelian beads.
References	:	Ambasta (1986), Kumar and Nagiyan (2006), Mackay (1933).

Capparis sepiaria Linn.

Syn: *C. glauca* Wall. *ex.* Hk. f. and Thoms.; *C. retusella* Thw.; *C. sepiaria* var. *incanescens* (DC.) Hk. f. and Thms.; *C. sepiaria* var. *retusella* (Thw.); *C. sepiaria* var. *vulgaris* Hk. f. and Thoms.; *C. spinosa* Linn.

Family: Capparidaceae

Eng., H. and Sans. Names	:	*Eng.*: Caper bush, Indian caper.
		H.: Ber, Hium, Kabra, Kanthari.
		Sans.: Ahimsa, Amlaphala, Dupravesha, Jali, Kakadani, Kanya, Vayastinduka.
Distribution	:	Rock and stony soil of N.W. India; throughout Panjab; Rajasthan–South.
Part Used	:	Root.
Reproductive Cycle	:	April–July.
Folk Use	:	Root piece stuck to the lock of hair of pregnant women for inducing abortion.
Active Constituents	:	Indole glucosinolates, glucobrassicin, neoglucobrassicin and 4–methoxy glucobrassicin characterized from roots.
Trade	:	The caper bush.
References	:	Bennet (1986), Khan and Khanum (2005), Punjani (2002), Rastogi and Mehrotra (1980-1984), Sharma (2003), Sood and Sudershna (2008).

Capsicum annuum Linn.

Syn: *C. frutescens* Cl.; *C. minimum* Roxb.

Family: Solanaceae

Eng., H. and Sans. Names	:	*Eng.*: Birds eye chilli, Red chilli, Red pepper.
		H.: Gachmirch, Lal mirch, Mattisa, Wangru.
		Sans.: Lanka, Marichi-phalam.
Distribution	:	Cultivated throughout the major parts of India.
Parts Used	:	Roots. Fruits.

Reproductive Cycle : May–November.

Folk Uses : Fruits tied around waist to keep off evil spirits. Nagas wear fruit as an ornament in ears and hair. A necklace of 7 pieces of its root tied to cure malaria.

Active Constituents : Active principles due to lutein, capsicin, capsaicin, phytofluene, solanine (fruits), saponin (capsicidin), 2 new phenolic amides–grossamide, and N-cis feruloyltyramine (roots).

Trade : Red chillies.

References : Anonymous (1950), Bennet (1986), Chauhan (1999), Hodson (1911), Jain (1991), Mill (1926), Molla and Roy (1996), Rastogi and Mehrotra (1960-1969, 1980-1984), Sharma (2003), Sood *et al.* (2005).

Cardiospermum halicacabum **Linn.**

Syn: *C. microcarpum* H.B.K.

Family: Sapindaceae

Eng., H. and Sans. Names : *Eng.*: Ballon vine heart seed, Black liquorice, Blister craper, Heart pea, Heart seeed, Winter cherry.

H.: Kanphata, Kanphuti.

Sans.: Indravalli, Karavi, Karnasphota, Nagna, Parvatagni, Tejovati.

Distribution : Wastelands, hedges and thickets throughout India.

Part Used : Seeds.

Reproductive Cycle : December–May.

Folk Use : A necklace prepared from seeds hung around the neck to prevent contagious diseases from affecting children.

References : Anonymous (1950), Khan and Khanum (2005), Punjani (2002), Sood and Sudershna (2008).

Careya arborea **Roxb.**

Syn: *C. orbiculata* Miers.; *C. sphaerica* Roxb.; *C. venelata* Oken.

Family: Barringtoniaceae

Eng., H. and Sans. Names : *Eng.*: Kumbi, Patana oak, Slow match tree, Wild guava.

H.: Kamba, Khumbi, Kumbh, Kumbi, Vakampa.

Sans.: Bhadrendrani, Kaidarya, Kalindi, Kumbhi, Patali, Pilu.

Distribution : Throughout greater parts of India.

Part Used : Stem.

Reproductive Cycle : March–June.

Folk Use : Stem fibres tied to the waist to cure 'dhuda' (a form of diarrhoea) in case of infants.

References : Jain *et al.* (1991), Khan and Khanum (2005), Kirtikar and Basu (1984), Sen and Behera (2007).

Carica papaya Linn.

Syn: *Papaya carica* Gaertner.

Family: Caricaceae

Eng., H. and Sans. Names : *Eng.*: Melon tree, Papaw, Paw-paw, Papaya.

H.: Anda, Kharbuja, Pappita.

Sans.: Brahmairandah, Chirbita, Eranda, Naikadala.

Distribution : Cultivated throughout India, naturalized throughout the warmer parts of India.

Part Used : Fruit

Reproductive Cycle : April–August.

Folk Use : Sliced raw fruit tied on wounds for healing by Jenukurubas.

Active Constituents : Fruits reported to possess β-carotene, cis-β-carotene, carotene, myristoleic, palmitoleic arachidic acids, pectins and free sugars.

Trade : Papaya.

References : Anonymous (1950), Khan and Khanum (2005), Kshirsagar and Singh (2007), Sharma (2003), Sood *et al.* (2009).

Carissa spinarum Linn.

Syn: *C. diffusa* Roxb.; *C. opaca* Stapf. *ex* Haines.

Family: Apocynaceae

Eng., H. and Sans. Names : *Eng.*: Karanda, Natal plum.

H.: Karaunda.

Sans.: Bokkarambuka, Jalipushpa, Karamadika, Vanalaya.

Distribution : Sub–Himalayan tract upto 1,000m; Trans–Indus territory and coasts of South Andamans.

Part Used : Flowers.

Folk Use	:	Flowers strung into garlands and put in their coiffures by women in the western coastal region and worn in hair.
References	:	Ambasta (1986), Jain (1991), Khan and Khanum (2005), Mehra *et al.* (1975), Watt (1889).

Caryota mitis **Lour.**

Family: Arecaceae

H. Name	:	*H.*: Mari-supari.
Distribution	:	Andaman Islands.
Part Used	:	Leaves.
Folk Use	:	Leaves used for making garlands.
References	:	Jain (1991).

Caryota urens **Linn.**

Family: Arecaceae

Eng., H. and Sans. Names	:	*Eng.*: Kittul, Sago, Toddy, Fish tail palm. *H.*: Mari. *Sans.*: Dirgha, Mada.
Distribution	:	Throughout India upto 1,200m.
Part Used	:	Seeds.
Reproductive Cycle	:	May–August.
Folk Use	:	Seeds used for making necklaces. Beads cut from nuts; plaited fibre collars worn by Hill Reddi men.
References	:	Mehra *et al.* (1975), Mittre (1997), Watt (1889).

Cascabela thevetia **(Linn.) Lipp.**

Syn: *Cerbera peruviana* Pers.; *C. thevetia* Linn.; *Thevetia neriifolia* Juss. *ex* Steud.; *T. peruviana* (Pers.) Merr.

Family: Apocynaceae

Eng., H. and Sans. Names	:	*Eng.*: Bastard oleander, Exile oleander, Lucky nut tree, Yellow oleander. *H.*: Pila, Pila kananir, Zard kaner. *Sans.*: Ashvamaraka, Nakharatlia, Sidhapushpa.
Distribution	:	Throughout India .
Parts Used	:	Flowers. Beads
Reproductive Cycle	:	July–September.

Folk Use	:	Flowers worn as garland by Bhils and other tribes. Beads strung and worn by tribes of West India.
Active Constituents	:	Flowers contain cardiac glycosides, thevetolin querectin–3–galactoside, 3–digalastoside, tamarixetin, and theretolin.
References	:	Bennet (1986), Chauhan (1991), Gogoi and Borthakur (1991), Jain (1991), Mehra *et al.* (1975), Khan and Khanum (2005), Rastogi and Mehrotra (1980-1984), Singh and Pandey (1998), Sood *et al.* (2009).

Casearia elliptica **Willd.**

Syn: *Anavinga lanceolata* Lam.; *C. ovata* Roxb.; *C. tomentosa* Roxb.

Family: Flacourtiaceae

H. and Sans. Names	:	*H.*: Bairi, Beri, Chilla. *Sans.*: Chilhaka, Kampillika.
Distribution	:	Moist hilly locations upto 900m from foothills of Himalaya southwards and Andaman Islands.
Part Used	:	Leaves.
Reproductive Cycle	:	March–May.
Folk Use	:	Leaves bandaged on the abdomen to cure pain, gastric disorders, external wounds and blisters.
Active Constituents	:	Leaves reported to possess hyperoside, arginine, aspartic acid, cysteine, glycine, leucine, proline, serine, phenylalanine and tryptophan.
Trade	:	Chilla. Churchu.
References	:	Bennet (1986), Chatterjee and Pakrashi (1997), Khan and Khanum (2005), Rastogi and Mehrotra (1980-1984), Singh and Pandey (1998), Sood and Sudershna (2008).

Casearia graveolens **Dalz.**

Family: Flacourtiaceae

H. Names	:	*H.*: Aloal, Chilla, Chilli, Kathera, Naro, Pimpri.
Distribution	:	Upper Gangetic Plains, westwards to Chenab, Garhwal, Kumaon, Sikkim, Mayanmar, Konkan, Deccan.
Part Used	:	Beads.
Reproductive Cycle	:	May–July.
Folk Use	:	Beads used for making ornaments.
References	:	Chandra (1997), Jain (1991), Kirtikar and Basu (1984).

Cassia auriculata **Linn.**

Syn: *Senna auriculata* Roxb.

Family: Fabaceae

Eng., H. and Sans. Names	:	*Eng.*: Avaram, Tanner's senna, The tanner's cassia. *H.*: Aswal, Tarval, Taval. *Sans.*: Avarttaki, Hemapushpam, Mayahari.
Distribution	:	Dry zones of Peninsular India, particularly common in A.P, Tamil Nadu and Karnataka.
Parts Used	:	Flowers. Young shoots.
Reproductive Cycle	:	January–July.
Folk Use	:	Flowers and young shoots tied over the stomach twice daily for 3 days to check colic pain even among adults.
Active Constituents	:	These are β-sitosterol and kaempferol (flowers). 3 new saturated higher fatty ketoalcohols and amodin (leaves).
Trade	:	Avaram.
References	:	Chandra (1997), Jain *et al.* (1991), Khan and Khanum (2005), Muthukumaraswamy *et al.* (2003), Rastogi and Mehrotra (1970-1979), Sood and Sudershna (2008).

Cassia fistula **Linn.**

Syn: *C. rhombifolia* Roxb.; *Cathartocarpus fistula* Pers.

Family: Fabaceae

Eng., H. and Sans. Names	:	*Eng.*: Drum stick, Golden shower, Indian laburnum, Purging cassia. *H.*: Amaltas, Kilvali, Kirala, Sinar. *Sans.*: Aragvadha, Chaturangula, Kritamalluka, Suvarnaka.
Distribution	:	Throughout India.
Parts Used	:	Flowers. Stem bark. Pods. Wood.
Reproductive Cycle	:	April–November.
Folk Uses	:	Bhils keep its dry pods under pillow of their children to stop bad dreams and protecting them from getting frightened while asleep. A necklace of its stem bark worn for checking malarial fever by Khonds and Porjas. Wood used in framework of bed (cot), for post delivery treatment.

Active Constituents	:	Isolated constituents are arginine, leucine, methionine, phenylalanine, tryptophan, aspartic acid, glutamic acid (fruits), emodin, chrysophanol, rhein, sennidin A and B (pods), steroids, glycosides, flavonoids, tannins, quinines, disaccharides, phlobaphenes (bark).
Trade	:	Fistula or Cassia.
References	:	Asolkar *et al.* (1992), Bhatt *et al.* (2003), Jadhav (2008), Jain (1994), Khan and Khanum (2005), Punjani (2002), Rastogi and Mehrotra (1980-1984), Singh and Pandey (1998), Sinha and Sinha (2001), Sood *et al.* (2005).

Cassia tora Linn.

Syn: *C. foetida* Salisb.; *C. gallinaria* Collad.; *C. humilis* Collad.; *C. obtusifolia* Linn.; *C. toroides* Roxb.

Family: Fabaceae

Eng., H. and Sans. Names	:	*Eng.*: Cofee weed, Foetid cassia, Ring worm plant, Sickle senna. *H.*: Chakramardah, Chakunda, Panevar. *Sans.*: Avudham, Chakramardah, Chakri.
Distribution	:	Throughout India.
Parts Used	:	Leaves. Roots.
Reproductive Cycle	:	August–October.
Folk Uses	:	A necklace of pieces of its roots cures children suffering from 'dhiulajawara' chronic fever with stiffness of muscles. Ring made from its roots worn in left finger and that from *Cassia occidentalis* in right hand for exorcism by Jenukurubas. Root tied on arm in case of severe one sided headache. A leaf tied to the left hand on Sunday or Tuesday to treat malarial fever.
Active Constituents	:	Principle constituents of leaves are d-manitol, myricyl alcohol, β-sitosterol, emodin whereas roots contain anthraquinones, β-sitosterol, helminthosporin, O-methyl chrysophanol, betulinic acid and 2,5-dimethoxybenzoquinone.
Trade	:	Sickle senna.
References	:	Asolkar *et al.* (1992), Chauhan (1999), Jain *et al.* (1991), Khan and Khanum (2005), Kshirsagar and Singh (2007), Rastogi and Mehrotra (1985-1989), Sahoo and Bahali (2003), Singh and Narain (2008), Sood *et al.* (2005).

Cassytha filiformis Linn.

Syn: *C. americana* Meissn.

Family: Lauraceae

Eng., H. and Sans. Names	:	*Eng.*: Doddar-laurel, Love vine. *H.*: Amarbel. *Sans.*: Akasvalli.
Distribution	:	Throughout the greater parts of India, particularly nearby the sea coast.
Part Used	:	Whole Plant.
Reproductive Cycle	:	Throughout the year.
Folk Use	:	A portion of plant tied by Santals around neck, arm and ankles to cure rickets.
Active Constituents	:	Plant contains cassythine, cassythidine and dulcitol.
References	:	Ambasta (1986), Anonymous (1950), Asolkar *et al.* (1992), Chatterjee and Pakrashi (1997), Dhiman (2003).

Catharanthus roseus (Linn.) G. Don.

Syn: *Lochnera rosea* (Linn.) Reichb.; *Vinca rosea* Linn.

Family: Apocynaceae

Eng., H. and Sans. Names	:	*Eng.*: Madagascar periwinkle, Old maid, Red periwinkle. *H.*: Sadabahar, Sada suhagan, Sada suhagi. *Sans.*: Nityakalyani, Rasna, Sadampushpa.
Distribution	:	Throughout India.
Part Used	:	Flowers.
Reproductive Cycle	:	Throughout the year.
Folk Use	:	Used as adornments.
Active Constituents	:	Flowers contain vinblastine, coronaridine, vindorosine, vindoline, catharanthine, mitraphlline, ajmalicine, tetrahydroalstonnie and reserpine.
Trade	:	Periwinkle.
References	:	Bennet (1986), Dhiman (2003), Gogoi and Borthakur (1991), Khan and Khanum (2005), Rastogi and Mehrotra (1970-1979), Sharma (2003), Sinha and Sinha (2001), Sood *et al.* (2009).

Caulerpa racemosa **(Forssk.) W. V. Bosse.**

Syn: *Caulerpa feldmannii* Rayss. and Edelstein.; *Fucus racemosa* Forssk.; *F. uvifer* Turner.

Family: Chlorophyceae

Eng. Name	:	*Eng.*: Grape caulerpa.
Distribution	:	Occurs abundantly in the Mediterranean sea.
Part Used	:	Whole Plant.
Folk Use	:	Used as a necklace.
Reference	:	Chakrabarty and Balkishnan (2003), Fritsch (1948).

Cayratia pedata **(Lamk.) Juss. *ex* Gagnep.**

Syn: *Cissus pedata* Lamk.; *Vitis pedata* Vahl. *ex* Wall.

Family: Vitaceae

H. and Sans. Names	:	*H.*: Goalalita. *Sans.*: Godhapadi, Suvaha.
Distribution	:	Bihar, W.B., Assam upto 900m.
Part Used	:	Root.
Reproductive Cycle	:	May–October
Folk Use	:	3-5 cm piece of its root tied as an amulet on the wrist for fever.
References	:	Ambasta (1986), Anonymous (1950), Bennet (1986), Chatterjee and Pakrashi (1997), Girach *et al.* (1996), Prajapati *et al.* (2006).

Celastrus paniculatus **Willd.**

Syn: *C. alnifolia* Don.; *C. dependens* Wall.; *C. multiflorus* Roxb.; *C. nutans* Roxb; *C. rothiana* Roem.; *Ceanothus paniculatus* Roth.; *Scutia paniculata* Don.

Family: Celastraceae

Eng., H. and Sans. Names	:	*Eng.*: Black oil plant, Climbing staff plant, Intellect tree, Staff tree. *H.*: Apang, Jamrasi, Kondgaidh, Malkagni, Pheng. *Sans.*: Avatapadi, Jyotishmati, Kanguni, Pitataila.
Distribution	:	Sub-Himalayan tract upto 2,000m, Hilly areas of Maharashtra, Gujarat, Tamil Nadu.
Part Used	:	Fruit.

Reproductive Cycle	:	Flowers at various times throughout the year, Seeds ripen in summer and autumn.
Folk Use	:	Garland made from the fruits used by women.
Trade	:	Jyotismatis.
References	:	Anonymous (1950), Chatterjee and Pakrashi (1997), Hosagoudar and Henry (1996a), Khan and Khanum (2005), Sharma (2003), Sood and Sudershna (2008).

***Celosia argentea* Linn.**

Syn: *C. argentea* Linn. var. *cristata* Kuntze.; *C. cristata* Linn.

Family: Amaranthaceae

Eng., H. and Sans. Names	:	*Eng.*: Cox comb plant, Quail grass, Wild cock's comb.
		H.: Debkoti, Safed murgha, Sarwari, Survari.
		Sans.: Pothura, Vitunna.
Distribution	:	A common weed throughout India.
Part Used	:	Flowers.
Reproductive Cycle	:	Flowers and Fruits during winter.
Folk Use	:	Flowers used as adornments.
References	:	Ambasta (1986), Bennet (1986), Chatterjee and Pakrashi (1997), Gogoi and Borthakur (1991), Khan and Khanum (2005), Sood and Sudershna (2008).

***Celtis cinnamomea* Lindl.**

Syn: *C. timorensis* Lindl. *ex* Planch.

Family: Ulmaceae

Distribution	:	Peninsular India.
Part Used	:	Bark.
Folk Use	:	Men of little Andaman wear girdles made from bark.
Reference	:	Radcliffe-Brown (1922).

***Ceriscoides turgida* (Roxb.) Tirvengadum.**

Syn: *Gardenia montana* Roxb.; *G. turgida* Roxb.

Family: Rubiaceae

H. and Sans. Names	:	*H.*: Khurtari, Thanella.
		Sans.: Karhata.
Distribution	:	Reported from sub-Himalayan tract: Garhwal–Bhutan, C. India, S. India.

Part Used	:	Roots.
Reproductive Cycle	:	April–June.
Folk Use	:	Natives of Chotta Nagpur wore the roots of this plant on their wrists as a charm.
Active Constituents	:	Roots contain gardenin A, B and E, oleanolic acid, α-amyrin, β-sitosterol and d-mannitol.
References	:	Asolkar *et al.* (1992), Chandra (1997), Dhiman (2003).

Chenopodium album Linn.

Syn: *C. giganteum* Don.; *C. laciniatium* Linn.; *C. nepalense* Hort.; *C. purpurascens* Ham.; *C. viride* Linn.

Family: Chenopodiaceae

Eng., H. and Sans. Names	:	*Eng.*: All good, Bacon weed, Biacon weed, Dirt weed, Dirty dick, Drought weed, Fat hen, Milds, Miles, Mutton-tops, Wild spinach. *H.*: Bathua, Bathusag, Chandanbedu. *Sans.*: Agralohita, Chakravarti, Chillika, Ghanaghana, Kankella, Mahaddala, Mridupatri, Panshupatra, Shakarata, Shakavira, Tuni, Vastu, Vastuka, Vastuki, Vasuka.
Distribution	:	Cultivated as a pot herb.
Part Used	:	Leaves.
Folk Use	:	Leaves tied to cure sprains.
Active Constituents	:	Isolated constituents from leaves are ascorbic acid, β-carotene, catechins, caffeic acid, p-coumaric acid, ferulic acid, β-sitosterol, xanthotoxin, n–triacontanol, scopoletin, spinasterol and 7-stigmasterol.
Trade	:	Wild spinach.
References	:	Chatterjee and Pakrashi (1997), Kirtikar and Basu (1984), Sharma (2003), Singh and Pandey (1998), Sood and Thakur (2004).

Chlorophytum arundinaceum Baker.

Syn: *Anthericum tuberosum* Herb.; *Ornithogalum indicum* Koenig.

Family: Liliaceae.

H. and Sans. Names	:	*H.*: Biskandri, Safed musli. *Sans.*: Musali, Sweta-musli.
Distribution	:	E. Himalaya, Assam, Bihar.

Part Used : Whole Plant.

Folk Use : There is a common belief that a contact with this plant can bring an errant wife back to her husband.

Trade : Safed musli.

References : Hembrom (1996), Jain *et al.* (1991), Sharma (2003), Sood *et al.* (2005).

Cirsium lipskyi **Petrak.**

Syn: *Cnicus griffithii* Hk. f.

Family: Asteraceae

Distribution : N. East India.

Part Used : Spine.

Folk Use : Used for tattooing.

Reference : Saklani and Jain (1994).

Cissampelos pareira **Linn.***

Syn: *C. convolvulacea* Willd.; *C. delicatula* Miers.; *C. discolor* Buch.; *C. diversa* Miers.; *C. elata* Miers.; *C. eriantha* Miers.; *C. grallatoria* Miers.; *C. hernandifolia* Wall.; *C. hirsuta* Buch.; *C. obtecta* Wall.; *C. orbiculata* Buch.

Family: Menispermaceae

Eng., H. and Sans. Names : *Eng.*: False periera, Periera brava, Velvet leaf.

H.: Akanudi, Dekhnirbissi, Harjeuri, Pari.

Sans.: Ambashthai-patha, Ruchaishya, Vidhakarni.

Distribution : Throughout India.

Parts Used : Stem, Leaves.

Reproductive Cycle : April–September.

Folk Uses : Symbolic wearing of pieces of its stem comprising of 7 nodes around neck as an amulet cures headache. Leaves tied to cure boils.

Trade : 'Patha'.

References : Aminuddin and Girach (1991), Bennet (1986), Jain *et al.* (1991), Khan and Khanum (2005), Rastogi and Mehrotra (1970-1979), Sen and Behera (2007), Sood *et al.* (2005).

Cissus quadrangularis Linn.

Syn: *C. edulis* Dalz.; *Vitis quadrangularis* Wall. *ex* Wight.

Family: Vitaceae

Eng., H. and Sans. Names	:	*Eng.*: Adamant creeper, Square stalked vine, The edible stemmed vine. *H.*: Hadjora, Harsankari. *Sans.*: Asthisamhara, Vajravalli.
Distribution	:	Commonly planted in gardens throughout the hotter parts of India.
Part Used	:	Stem.
Reproductive Cycle	:	May–August.
Folk Uses	:	To cure backache, the patient advised to take rest on bed spread overwith it's stem pieces for at least 2 hours daily for 3 days. Also used for making ornaments.
Active Constituents	:	Stem known to contain 2 tetracyclic triterpenoids and 2 steroidal principles.
Trade	:	Hatjora.
References	:	Anonymous (1950), Bennet (1986), Chatterjee and Pakrashi (1997), Khan and Khanum (2005), Khare (2004), Kumbhojkar *et al.* (1991), Punjani (2004), Sharma (2003).

Citrus maxima (Burm.) Merr.

Syn: *Aurantium maximum* Burm.; *C. decumana* Linn.; *C. grandis* (Linn.) Osbeck.

Family: Rutaceae

Eng., H. and Sans. Names	:	*Eng.*: Forbidden fruit, Paradies apple, Pommel-moose shaddock, Pummelo. *H.*: Batavinebu, Botornebu, Chakotra. *Sans.*: Madhukarkati.
Distribution	:	Cultivated in the plains.
Part Used	:	Fruit.
Reproductive Cycle	:	February–November.
Folk Use	:	Flowers used as adornments.
Active Constituents	:	5-geranyloxy-7-methoxycoumarin, and isoimperatorin, bergapten an 5-geranyloxypsoralen isolated from fruits.
References	:	Asolkar *et al.* (1992), Bennet (1986), Gogoi and Borthakur (1991), Rastogi and Mehrotra (1960-1969, 1990-1994), Sood *et al.* (2009).

***Citrus medica* Linn.**

Syn: *C. aurantifolia* (Christm.) Swingle.; *C. limetta* Risso.; *C. limonum* Risso.; *C. medica* Linn. var. *acida* Watt.

Family: Rutaceae

Eng., H. and Sans. Names	:	*Eng.*: Adam's apple, Cedrat, Citron, Lemon, Melon lime. *H.*: Bara nimbu, Kagzi nimbu. *Sans.*: Amlakshera, Bijaphalaka, Jambira.
Distribution	:	Tropical parts of India including valleys of the Himalaya and Khasi Hills.
Parts Used	:	Leaf, Flowers, Fruit.
Reproductive Cycle	:	April–September.
Folk Uses	:	Fruits used as adornment. Leaves used for hair dressing. Flowers worn in hair since medieval period and fruit strung as a charm.
Active Constituents	:	Fruit peel contains apigenin, luteolin, chrysoseriol, quercitin, isorhamnetin, citrusins A, B, C, D, scopoletin, umbelliferone, hesperidin.
Trade	:	Sour lime.
References	:	Dhiman (2003), Dymock (1889a), Gogoi and Borthakur (1991), Saklani and Jain (1994), Sharma (2003), Sood *et al.* (2009).

***Citrus reticulata* Blanco.**

Syn: *C. aurantium* Linn.; *C. chrysocarpa* Lushington.; *C. nobilis* Lour.

Family: Rutaceae

Eng., H. and Sans. Names	:	*Eng.*: Loose skinned orange, Mandarin or Tangerine orange. *H.*: Santara. *Sans.*: Narangah.
Distribution	:	Widely cultivated in sub-tropical regions of India.
Part Used	:	Fruit.
Reproductive Cycle	:	January–October.
Folk Use	:	As adornment.
Active Constituents	:	These are β-pinene, dl-limonene, linalool, citropten, tangeretin, citromitin, hesperidin, β-sitosterol, β-myrin (fruit).
Trade	:	Bergamot orange.

References : Chatterjee and Pakrashi (1997), Gogoi and Borthakur (1991), Khan and Khanum (2005), Rastogi and Mehrotra (1990-1994), Sharma (2003).

Claoxylon indicum (Bl.) Hassk.

Syn: *C. polot* (Burm.f.) Merr.

Family: Euphorbiaceae

Eng. Name : *Eng.*: Indian claoxylon.

Distribution : S. India.

Part Used : Leaves.

Folk Use : Leaves kept below head while sleeping to check giddiness.

References : Ambasta (1986), Dagar and Dagar (1996).

Cleome viscosa Linn.

Syn: *C. icosandra* Linn.; *Polanisia icosandra* Wight. and Arn.

Family: Capparidaceae

Eng., H. and Sans. Names : *Eng.*: Sticky cleome viscid herb, Wild mustard.

H.: Boogra, Hulhul, Jangli harsar, Kamphutia.

Sans.: Adityabhakta, Arkakanta, Barbara, Manduki, Raviprita, Ravishta.

Distribution : Throughout tropical and sub-tropical India, rocky sandy soils, cultivated fields, roadsides.

Part Used : Leaves.

Reproductive Cycle : Throughout year.

Folk Use : Leaves tied with wet white cloth on head to treat headache.

References : Ambasta (1986), Anonymous (1950), Chatterjee and Pakrashi (1997), Dhiman (2007), Khan and Khanum (2005), Kumar and Pullaiah (1998).

Clerodendrum indicum (Linn.) O. Kuntze.

Syn: *C. siphonanthus* R. Br.; *Siphonanthus indica* Linn.

Family: Verbenaceae

Eng., H. and Sans. Names : *Eng.*: Tube flower, Turk's turban.

H.: Barangi, Bharangi.

Sans.: Bharangi, Brahmanayashtika.

Distribution	:	Wasteplaces, hedges, roadsides throughout the warmer parts of India.
Part Used	:	Branches.
Reproductive Cycle	:	May–October.
Folk Use	:	A necklace of beads of its branches put on as a charm on necks of the children suffering from skin diseases.
Active Constituents	:	(24 S)-Ethyl cholesta-5, 22,25-trien-3β-ol and β-sitosterol are the active principles of stems.
References	:	Ambasta (1986), Anonymous (1950), Bennet (1986), Dhiman (2003), Rastogi and Mehrotra (1980-1984).

Clerodendrum serratum **(Linn.) Moon.**

Syn: *C. fortunatum* Burm. *ex* Moldenke.; *C. serrulatum* Spreng *ex* Razi.; *Volkameria serrata* Linn.

Family: Verbenaceae

Eng., H. and Sans. Names	:	*Eng.*: Beetle killer, Blue-flowered glory tree. *H.*: Barangi, Bharangi, Duma, Gant-baharangi. *Sans.*: Bharangi, Bharang, Bhargi.
Distribution	:	South and E. India, especially in W. Bengal.
Parts Used	:	Leaf. Root.
Reproductive Cycle	:	Flowering (July-August), Fruiting cold season.
Folk Uses	:	Leaf tied on belly for indigestion. Root piece contained in a small brass container used as a talisman on the left arm of the woman and right arm of man with a view that poisonous snakes do not bite such a talisman bearer.
Active Constituents	:	These are sapogenin, triterpene, starch, resin, glucose, D-(-) mannitol (roots) and a bitter principle (leaves).
Trade	:	Bhargi.
References	:	Chauhan (1999), Khan and Khanum (2005), Sahoo and Bahali (2003), Saklani and Jain (1994), Sharma (2003), Sood and Sudershna (2008), Sood *et al.* (2005).

Clerodendrum thomsonae **Balf.**

Family: Verbenaceae

Eng. Names	:	*Eng.*: Bleeding heart, Bleeding heart vine, Bleeding heart glorybower.
Distribution	:	Throughout N. India.
Part Used	:	Flowers.

Reproductive Cycle	:	Flowering in April or May.
Folk Use	:	As adornments.
References	:	Gogoi and Borthakur (1991).

Clerodendrum viscosum Vent.

Syn: *C. infortunatum* non Linn.

Family: Verbenaceae

H. and Sans. Names	:	*H.*: Bhant, Thuner. *Sans.*: Barhibhra, Kukkura, Simaroma, Sukabarha, Sukapushpa.
Distribution	:	W. Ghats throughout India.
Part Used	:	Flowers.
Reproductive Cycle	:	October–December.
Folk Use	:	Flowers worn for worship in Bastar district.
References	:	Jain (1963).

Clitoria ternatea Linn.

Syn: *Terneata vulgaris* H. B. K.

Family: Fabaceae

Eng., H. and Sans. Names	:	*Eng.*: Butterfly pea, Mezereon, Winged leaved clitoria. *H.*: Aprajita, Kajina, Kalina, Kalizer, Kowa, Shobanjan. *Sans.*: Aprajita, Bhadra, Gavakshi, Kinihi, Koyal, Romavalli, Shveta, Suputri, Vishnukantri.
Distribution	:	Throughout India.
Parts Used	:	Flowers. Roots. Whole Plant.
Reproductive Cycle	:	August–October.
Folk Uses	:	Flowers used as adornments. A small piece of root tied to the right ear of the patient to check headache. Climber along with leaves and flowers tied around the neck of cattle to cure swelling of skin by Soligas.
Active Constituents	:	Active properties due to malvidin-3β-glucoside and 3-methyl ether of delphinidin-3-β-glucoside (flowers) and taragerol (roots).
Trade	:	Aparajit.
References	:	Gogoi and Borthakur (1991), Jain *et al.* (1991), Khan and Khanum (2005), Khare (2004), Kshirsagar and Singh (2007), Rastogi and Mehrotra (1970-1979, 1980-1984, 1985-1989), Sen and Behera (2007).

Cochlospermum religiosum (Linn.) Alston.

Syn: *C. gossypium* DC.; *C. religiosum* Linn.

Family: Cochlospermaceae

Eng., H. and Sans. Names	:	*Eng.*: Silk cotton tree. *H.*: Galgal, Kumbi. *Sans.*: Girisalmalika, Silakarpasika.
Distribution	:	Drier regions of India; less common in W. India.
Part Used	:	Flowers.
Folk Use	:	Flowers worn in Bastar district for worship.
References	:	Ambasta (1984), Jain (1963).

Cocos nucifera Linn.

Syn: *C. nana* Griff.

Family: Arecaceae

Eng., H. and Sans. Names	:	*Eng.*: Cocoa nut, Coconut palm, Coconut tree. *H.*: Narel, Nariel, Nariyal, Nariyal kaper. *Sans.*: Dakshinatriya, Dridanira, Garikera, Jataphala, Junga, Langali, Mangalya, Mutkuna, Narikari, Putodaka, Rasaphala, Sadapushpa, Varaphala.
Distribution	:	Cultivated in Karnataka, Kerala and Tamil Nadu, also found wild in S. India upto 1,000 m above sea level.
Parts Used	:	Fruit shell, Kernels, Fiber, Leaves.
Folk Uses	:	Fruit shells cut into rings by Yanadi men, bangles by Kunbis and Gadulias. Munda boys make belts from fiber, and Nicobarese make leaf collars. Dried kernels of fruit cut into garlands which Hindu women wear on religious ceremonies.
Active Constituents	:	Shell contains crotonaldehyde, furfural and acetic acid.
Trade	:	Coconut.
References	:	Anonymous (1950), Dhiman (2003), Rastogi and Mehrotra (1970-1979), Sastry (1966), Singh and Pandey (1998).

Codiaeum variegatum Blume

Family: Euphorbiaceae

Eng. Name	:	*Eng.*: Croton.
Distribution	:	Cultivated only, from sea level to mid-montane.

Parts Used : Leaves. Shoots.

Folk Use : Used as adornments.

Active Constituents : Leaves rich in polyphenols, chlorogenic, protocatechuic, p-hydroxybenzoic, cis, and trans ferulic and vanillic acids.

References : Asolkar *et al.* (1992), Gogoi and Borthakur (1991), Prajapati *et al.* (2006), Rastogi and Mehrotra (1985-1989).

Coelogyne punctulata **Lindl.**

Syn: *C. nitida* Lindl.; *Cymbidium nitidum* Roxb.; *C. ochracea* Lindl.

Family: Orchidaceae

Distribution : Himalaya, (Kumaon–Bhutan): 1,300-2,400m.

Part Used : Flowers.

Folk Use : Its white flowers used by women folk in Assam for their personal adornments.

References : Bennet (1986), Rastogi and Mehrotra (1980-1984), Sood *et al.* (2006).

Coix gigantea **Koen. *ex* Roxb.**

Family: Poaceae

Eng. Name : *Eng.*: Wild adlay.

Distribution : Marshy places in India.

Part Used : Fruit cases.

Folk Use : Fruit cases used for making beads and rosaries which are used as adornments in the eastern hill districts of India.

References : Anonymous (1950), Mehra *et al.* (1975).

Coix lacryma-jobi **Linn.**

Syn: *C. arundinacea* Lamk.; *C. lachryma* Linn.

Family: Poaceae

Eng., H. and Sans. Names : *Eng.*: Job's tears.

H.: Baru, Dabhir, Ganduta, Garahadua, Garun, Gurlu, Kaiya, Kasei, Sankhru.

Sans.: Gavedhukah, Gavedu, Gojivha, Jagradi.

Distribution : Meghalaya and Sikkim, ascending to 500m in hills.

Parts Used	:	Bracts. Seeds.
Reproductive Cycle	:	September–February
Folk Uses	:	Beads used for making necklaces in Manipur and worn by Shompen, Mundas, Bondos, Karans and Kadirs men and women. Hard bracts also used to prepare necklaces to bride and bridegroom in weddings. Angami Naga tribes prepare elegant ear rings in which a rosette of seeds surrounds the wings of a greenish beetle.
Active Constituents	:	An anticancer agent coixenolide, 2-hydroxy-7-methoxy-2H-1, 4-benzoxazin-3-one and 6-methoxybenzoxazolinone isolated from seeds.
References	:	Balfour (1871), Chakrabarty and Balakrishnan (2003), Diwakar and Ansari (2003), Jain (1991), Khan and Khanum (2005), Mehra *et al.* (1975), Roy (1925), Saklani and Jain (1994), Singh (2000), Watt (1889).

***Coix lacryma-jobi* Linn. var. *stenocarpa* Oliv.**

Family: Poaceae

Distribution	:	Upper Assam, Karen hills (Eastern Myanmar)
Part Used	:	Fruit cases.
Folk Use	:	Necklaces and ear rings made from fruit cases used by tribes in Assam and N. eastern areas.
References	:	Hooker (1897), Mehra *et al.* (1975).

***Commelina benghalensis* Linn.**

Syn: *C. canescens* Vahl.; *C. cucullata* Linn.; *C. deliculata* Sch.; *C. mollis* Jacq.; *C. nervosa* Burm.; *C prostrata* Regel.

Family: Commelinaceae

Eng., H. and Sans. Names	:	*Eng.*: Bengal day flower. *H.*: Kanchara, Kanchera. *Sans.*: Kancata, Marishajalaja, Paniya, Tanduliya.
Distribution	:	Moist regions throughout India.
Part Used	:	Flowers.
Reproductive Cycle	:	August–December.
Folk Use	:	Used as adornments.
References	:	Ambasta (1986), Anonymous (1950), Gogoi and Borthakur (1991), Khan and Khanum (2005).

Congea tomentosa Roxb.

Syn: *C. tomentosa* var. *azurea* (Wall. *ex* Walp.) Cl.; *C. villosa* (Roxb.) Cl.

Family: Verbenaceae

Eng. Names : *Eng.*: Lavender Wreath, Shower of orchids, Wooly congea.

Distribution : Assam. Manipur. Tamil Nadu. West Bengal.

Part Used : Stem.

Folk Use : Stem made into a ring and tied to the neck of domestic animals (cattle, pig, goat) to get rid of sore worms.

References : Bennet (1986), Lalramnghinglova (2003).

Corchorus capsularis Linn.

Syn: *C. marua* Ham.

Family: Tiliaceae

Eng., H. and Sans. Names : *Eng.*: Jute plant, White plant.

H.: Narcha, Titapat.

Sans.: Kalasaka.

Distribution : Hotter parts of India.

Part Used : Root.

Reproductive Cycle : July–October.

Folk Use : Root collected in early morning on Sunday and tied to the waist to stop baby from crying.

Active Constituents : Properties of roots attributed to triterpenoids, corosin and β-sitosterol.

References : Ambasta (1986), Asolkar *et al.* (1992), Chatterjee and Pakrashi (1997), Kirtikar and Basu (1984), Rastogi and Mehrotra (1970-1979, 1980-1984), Sen and Behera (2007).

Cordia gharaf (Forssk.) Ehrenb. *ex* Asch.

Syn: *C. angustifolia* Roxb.; *C. rothii* Roem. and Schult.; *C. sinensis* Lamk.; *Cornus gharaf* Forsk.

Family: Ehretiaceae

Eng., H. and Sans. Names : *Eng.*: Sebestan.

H.: Gondi, Gondni, Gundi.

Sans.: Laghushleshmataka.

Distribution : Panjab, Sind, Rajasthan, Gujarat.

Part Used	:	Stem bark.
Reproductive Cycle	:	March–June.
Folk Use	:	Ladies and children chew the stem bark with 'kattha' as a substitute of betel leaf to redden their lips and tongue at the occasion of festival and marriage.
Active Constituents	:	Stem contains sitosterol.
References	:	Ambasta (1986), Anonymous (1950), Bennet (1986), Asolkar *et al.* (1992), Rastogi and Mehrotra (1970-1979), Singh and Pandey (1998).

Corypha umbraculifera Linn.

Family: Arecaceae

Eng. and Sans. Names	:	*Eng.*: Fan palm, Great fan palm, Holy palm, Java fan palm, Malabar coast fan palm, South Indian talipot palm, Umbrella palm. *Sans.*: Alpayushi, Karalika, Katakali, Pakti, Sritala, Tali.
Distribution	:	Andamans, cultivated in tropical India.
Part Used	:	Seeds.
Folk Use	:	Polished seeds used for making buttons and small carved ornaments and as beads by Hindu devotees. Vegetable ivory from nuts cut into beggars beads.
References	:	Ambasta (1986), Anonymous (1950), Balfour (1871), Jain (1991), Kirtikar and Basu (1984), Mehra *et al.* (1975), Wealth (1950).

Crataeva magna (Lour.) DC.

Syn: *Capparis magna* Lour.; *C. lophosperma* Kurz; *C. nurvala* Buch. Ham.; *C. religiosa* var. *nurvala* (Buch-Ham.) Hk. f. and Thoms.

Family: Capparidaceae

Eng., H. and Sans. Names	:	*Eng.*: Three leaves caper. *H.*: Barna, Barun, Bila, Bilasi, Varvunna. *Sans.*: Ajapa, Asmarighna, Barhapushpa, Kumara, Mahakapittha, Sadhuvriksha, Varunah.
Distribution	:	Throughout India.
Part Used	:	Leaves, Wood.
Reproductive Cycle	:	February–May.
Folk Uses	:	A leaf kept on the ear of the mother is believed to expel out the retained placenta. Amulets for adornments also cut from wood.

Active Constituents	:	Leaves contain l-stachydrine.
Trade	:	Varuna.
References	:	Anonymous (1950), Asolkar *et al.* (1992), Bennet (1986), Dhiman (2007), Jain *et al.* (1991), Khan and Khanum (2005), Sen and Behera (2007).

***Crinum asiaticum* Linn.**

Syn: *C. toxicarium* Roxb.

Family: Amaryllidaceae

Eng., H. and Sans. Names	:	*Eng.*: Poison bulb, St. John's lily. *H.*: Chindar, Kanmu, Kanwal, Pindar. *Sans.*: Bala, Durdharsha, Dusaha, Jambu, Malaghni, Nagapatra, Raktapushpi, Shrikanda, Viphala, Vishari, Vrikka.
Distribution	:	Throughout tropical India.
Part Used	:	Flowers.
Folk Use	:	Used as adornments to decorate their hairs.
Active Constituents	:	Active properties of flowering bulbs attributed to 2-phenanthri-done alkaloids-crinasiatine and crinasiadine.
References	:	Anonymous (1950), Bhargava (1997), Gogoi and Borthakur (1991), Kirtikar and Basu (1984), Rastogi and Mehrotra (1985-1989).

***Crinum latifolium* Linn.**

Syn: *C. zeylanicum* Linn.

Family: Amaryllidaceae

Eng. and Sans. Names	:	*Eng.*: Wide leaved crinum. *Sans.*: Madhuparnika, Vrishakarni.
Distribution	:	Wild, also cultivated as an ornamental .
Part Used	:	Flowers.
Folk Use	:	Flowers used by Onge women to adorn their hairdo.
Active Constituents	:	Chief principles due to glucoalkaloid–latisoline, 2 new epimeric pyrrolophenanthridine alkaloids, 2-epilycorine and 2-epipancrassidine in flowers.
References	:	Anonymous (1950), Chakrabarty and Balakrishnan (2003), Rastogi and Mehrotra (1980-1984, 1990-1994).

Crossandra undulaefolia Salisb.

Syn: *C. infundibuliformis* (Linn.) Nees.

Family: Acanthaceae

Eng. Names : *Eng.*: Firecracker flower, Firecracker plant, Yellow crossandra.

Distribution : Cultivated in Southern states of Karnataka, Tamil Nadu and Andhra Pradesh.

Part Used : Flowers.

Folk Use : Garlands of flowers used in Bombay.

References : Ambasta (1986), Namboothiry (1980).

Crotalaria spectabilis Roth.

Syn: *C. leschenaultii* DC.; *C. sericea* Retz.

Family: Fabaceae

H. and Sans. Names : *H.*: Jhunjhunia.

Sans.: Ghantarava.

Distribution : Tropical regions, throughout India proper, ascending to 1,000m in Kumaon.

Part Used : Fruits.

Reproductive Cycle : September–December.

Folk Use : Dry fruits with seeds tied on the wrist and feet by 'Bhil'girls as a substitute of metallic 'ghunghroos'.

Active Constituents : Active constituents of seeds are α-carotene, triacontane, 22,23-dihydro stigmasterol and sericine.

References : Ambasta (1986), Anonymous (1950), Bennet (1986), Hooker (1897), Kumar and Nagiyan (2006), Rastogi and Mehrotra (1960-1969), Singh and Pandey (1998).

Croton caudatus Geisel.

Family: Euphorbiaceae

Distribution : Eastern Himalaya, Assam, Bengal.

Part Used : Green stems.

Folk Use : Gum exuded after burning the green stems of *Croton caulatus*, *Lasianthus sikkimensis*, *Murraya paniculata* and *M. koenigii* applied for painting the teeth black

Active Constituents : Stem contains dotriacotanol, β-amyrin and β-sitosterol.

References : Asolkar *et al.* (1992), Borthakur (1997).

Croton roxburghii **Baldkr.**

Syn: *C. joufra* Roxb.; *C. laevigatus* Vahl.; *C. oblongifolius* Roxb.

Family: Euphorbiaceae

H. and Sans. Names	:	*H.*: Chucka.
		Sans.: Bhutamkusam, Naagadanti, Putrashreni.
Distribution	:	N. E. India, Chota Nagpur, M.P., Scrub forests in Penninsular India.
Part Used	:	Roots.
Reproductive Cycle	:	January–May.
Folk Use	:	An amulet of a piece of its root is tied on arm to ward off evil spirits.
Active Constituents	:	β-sitosterol, a diterpene alcohol, diterpene–oblongifoliol.
References	:	Bennet (1986), Chatterjee and Pakrashi (1997), Khan and Khanum (2005), Rastogi and Mehrotra (1960-1969, 1970-1979), Sood *et al.* (2005).

Curculigo capitulata **(Lour.) O.K.**

Syn: *C. recurvata* Dry.; *Leucojum capitulatum* Lour.; *Molineria capitulata* (Lour.) Herb.

Family: Amaryllidaceae

Eng. Name	:	*Eng.*: Large leaved curculigo.
Distribution	:	Himalaya from Nepal eastwards and Bengal alt. 300-1,000m.
Part Used	:	Leaf.
Folk Use	:	Leaf used for tying the lame leg of hens.
References	:	Anonymous (1950), Bennet (1986), Rastogi and Mehrotra (1990-1994), Saklani and Jain (1994).

Curcuma longa **Linn.**

Syn: *Amomum curcuma* Jack.; *Curcuma domestica* (Medik.) Val.

Family: Zingiberaceae

Eng., H. and Sans. Names	:	*Eng.*: Indian saffron, Turmeric.
		H.: Halad, Halda, Haldi, Halud, Hardi.
		Sans.: Aneshta, Bahula, Dirgharaga, Gauri, Haridra, Harita, Jayanti, Kanchani, Kaveri, Lakshmi, Mangalya, Nisha, Pavitra, Pinja, Shifa, Vara, Varnavati, Yamini.
Distribution	:	Widely in most parts of India.

Part Used : Dried Rhizome.

Reproductive Cycle : May–August.

Folk Use : A necklace of rhizome worn by the victim with chanting 'mantras' for early cure of jaundice in Banswara district of Rajasthan. Groom ties root on bride.

Active Constituents : Constituents isolated from rhizome are curcumin, diferuloylmethane, feruloyl-p-coumaroylmethane and di–p–coumaroylmethane.

References : Ambasta (1986), Anonymous (1950), Asolkar *et al.* (1992), Bennet (1986), Dymock (1889b), Kirtikar and Basu (1984), Rastogi and Mehrotra (1960-1985), Singh and Pandey (1998), Sood *et al.* (2005).

Curcuma neilgherrensis **Wight.**

Syn: *C. angustifolia* Dalz.

Family: Zingiberaceae

Eng., H. and Sans. Names : *Eng.*: East India Arrowroot, Narrow-leaved turmeric, Wild arrowroot.

H.: Tavakhira, Tikhur.

Sans.: Gavayodbhava, Godhumaja, Payakshira, Pishtika, Talakshira, Talasambhuta, Tandulodbhava, Tavakshira, Yavaja.

Distribution : Konkan, abundant on the S.W. slope of Nilgiris.

Part Used : Flowers.

Reproductive Cycle : Flowering in May.

Folk Use : Flowers worn by the women.

References : Ambasta (1986), Hosagoudar and Henry (1996a), Kirtikar and Basu (1984), Sood *et al.* (2005), Varghese (1996).

Curcuma zedoaria **(Christ.) Rosc.**

Syn: *Amomum zedoaria* Christ.; *Curcuma zerumbent* Roxb.

Family: Zingiberaceae

Eng., H. and Sans. Names : *Eng.*: Zedoary.

H.: Kachura, Kalihaldi.

Sans.: Dravida, Durlabha, Gandhamulaka, Jatala, Kalpaka, Mukhya, Sati, Vedhya.

Distribution : Cultivated throughout India upto 1,500m.

Part Used	:	Flowers.
Reproductive Cycle	:	April–May.
Folk Use	:	As adornments.
Trade	:	Zedoary.
References	:	Ambasta (1986), Gogoi and Borthakur (1991), Sood and Sudershna (2008).

Cynodon dactylon **(Linn.) Pers.**

Syn: *Agrostis linearis* Retz.; *C. linearis* Willd.; *C. stellstus* Willd.; *Digitaria dactylon* Scop.; *Panicum dactylon* Linn.; *Paspalum dactylon* DC.

Family: Poaceae

Eng., H. and Sans. Names	:	*Eng.*: Bahama grass, Bermuda grass, Creeping panic grass, Couch grass, Devil's grass, Doob grass, Doorwa. *H.*: Dhoboghas, Dub, Durba, Huryali, Ramghas. *Sans.*: Amari, Asitalata, Bahuvirya, Bharghvi, Durva, Gauri, Jaya, Mangala.
Distribution	:	Throughout India upto 3,000m.
Part Used	:	Whole Plant.
Reproductive Cycle	:	August–November.
Folk Use	:	During 'Gangaur' festival in N.W. Rajasthan, the unmarried tribal girls put the grass on their heads with the belief that they will get a husband of their choices. Women wear rings of culms for worship.
Trade	:	Doob-grass.
References	:	Anonymous (1950), Mookerji (1930), Rastogi and Mehrotra (1990-1994), Sharma (2003), Singh and Pandey (1998), Sood and Sudershna (2008).

Dalbergia sissoo **Roxb. *ex* DC.**

Syn: *D. pendula* Temore.

Family: Fabaceae

Eng., H. and Sans. Names	:	*Eng.*: Sissoo, South Indian Red wood. *H.*: Shisham, Sissai, Sissu. *Sans.*: Agurushinshupa, Mandalapatri, Shinshapa.
Distribution	:	W. Himalaya (Sikkim–Upper Assam), extensively planted along roadsides in different parts of India.
Part Used	:	Seeds.

Reproductive Cycle	:	May–August.
Folk Use	:	Green seeds of this deciduous tree worn by Santhal tribe girls as pendants for the ears. Marriage amulet cut from wood.
Active Constituents	:	Constituents isolated from seeds are isocaviunin, tectorigenin, dalbergin, biochanin A and 7-Hydroxy–4-methylcoumarin.
Trade	:	Shisham, Sissoo.
References	:	Anonymous (1952), Dhiman (2003), Jain (1991), Kirtikar and Basu (1984), Mehra *et al.* (1975), Rastogi and Mehrotra (1980-1984), Sood *et al.* (2009), Watt (1889, 1972)

Daphne papyracea **Wall.** ***ex*** **Steud.**

Syn: *D. bholua* Don.; *D. cannabina* Lour. *ex* Wall.; *D. odora* Don.; *D. papyrifera* Ham.

Family: Thymelaeaceae

Eng. and H. Names	:	*Eng.*: Himalayan mezercon, Nepal paper plant.
		H.: Bhancoi, Dunkotah, Gande, Kaghuti, Satpura, Setburusa, Setburwa.
Distribution	:	Found in Montane Himalaya from (Kashmir–Bhutan) and S. Tibet: 1,500-2,400m.
Parts Used	:	Leaf. Flowers.
Reproductive Cycle	:	April–October.
Folk Uses	:	Leaves used for making garlands. Garlands made from flowers used as adornments by N. E. region and Himalayan people.
Active Constituents	:	Leaves rich in daphnetin, daphenetin-8-α-glucoside, taraxerol and sitosterol-D-glucoside.
References	:	Bennet (1986), Dhiman (2003), Jain (1991), Mehra *et al.* (1975), Sood *et al.* (2009), Watt (1889, 1972).

Datura fastuosa **Linn. var.** ***alba*** **(Nees.) C.B. Clarke.**

Syn: *D. alba* Nees.; *D. metel* Linn.

Family: Solanaceae

Eng., H. and Sans. Names	:	*Eng.*: Downy datura, Thorn apple, Trumpet flower.
		H.: Dhattura, Kaladhatura, Sadahdhatura.
		Sans.: Dattur, Dhastura, Kanak, Kanakonmatta, Shivapriya, Unmatta.
Distribution	:	Throughout India.

Parts Used : Leaves. Root.

Reproductive Cycle : April–September.

Folk Uses : Leaves tied over the body for 2-3 days to cure myalgia. A necklace of 7-21 pieces of its root tied with a sacred thread for a week to cure typhoid or to the waist of pregnant woman to prevent miscarriage or abortion and allowed to remain there still delivery.

Active Constituents : Properties attributed to atropine, small amount of hyascyamine (leaves), hyoscyamine, scopolamine, ditigloyloxytropane, tigloidine, apohyoscine, meteloidine, tropine, 3-α-acetoxytropane, 3α, 6β-ditigloyloxytropane (roots).

Trade : Dhaturaa-bija, Dhaturaa-patti.

References : Anonymous (1952), Bennet (1986), Chauhan (1999), Jadhav (2008), Jain *et al.* (1991), Punjani (2002), Sen and Batra (1997), Sood and Sudershna (2008), Sood *et al.* (2009).

Datura innoxia* Mill.

Family: Solanaceae

Eng., H. and Sans. Names : *Eng.*: Thorn apple.

H.: Lal dhatura, Safed dhatura.

Sans.: Dhattura, Dhustura.

Distribution : N.W. Himalaya, Temperate regions of India, W. Ghats.

Parts Used : Leaves. Roots.

Reproductive Cycle : Throughout the year.

Folk Uses : Fresh leaves tied on wounds or swollen body parts for early cure. Roots kept in the navel of an expecting woman for easy delivery.

Active Constituents : Properties due to scopolamine (leaves), tropine, meteloidine, hyoscyamine, 6-β-bistigloyloxytropane, and 7-hydroxy-3,6-bis-tigloyioxytropane (roots).

References : Anonymous (1952), Katewa *et al.* (2001), Rastogi and Mehrotra (1970-1985), Singh and Pandey (1998), Sood and Sudershna (2008), Sood *et al.* (2005).

***Delonix regia* (Boj. *ex* Hk.) Rafin.**

Syn: *Poinciana regia* Boj. *ex* Hk.

Family: Fabaceae

Eng. and H. Names : *Eng.*: Flamboyant, Flame tree, Gold mohar.

H.: Gulmohar.

Distribution : Planted as avenue tree in gardens and roadsides throughout India.

Part Used : Flowers.

Folk Use : Showy scarlet red flowers used for adornments by the tribals in the form of garlands in W. India.

Active Constituents : Flowers found to have hentriacontane, quercetin, protocathechuic acid, and β-sitosterol.

References : Anonymous (1952), Asolkar *et al.* (1992), Awasthi and Goel (2000), Bennet (1986), Rastogi and Mehrotra (1960-1979), Singh and Pandey (1998).

Dendrobium crumenatum **Sw.**

Syn: *Angraecum crumenatum* Rhump.

Family: Orchidaceae

Eng. Names : *Eng.*: Bag shaped dendrobium, Sparrow orchid, Tropical pegion orchid.

Distribution : Semi deciduous lowland forest in India and Sri Lanka.

Part Used : Stems.

Reproductive Cycle : Flowering (March–July).

Folk Use : Stems used to ornament men's belts, in Andaman Islands.

References : Radcliffe-Brown (1922).

Dendrobium densiflorum **Wall. f.**

Syn: *D. clavatum* Wall. *ex* Lindl.; *D. thyrsiflorum* Reichb. f.

Family: Orchidaceae

Eng. Name : *Eng.*: Pineapple orchid.

Distribution : N. E States of India.

Parts Used : Stem bark. Flowers.

Reproductive Cycle : Flowering (August–September).

Folk Uses : Stem bark used as earing. Flowers worn by Naga women of Manipur behind the ears.

References : Arditi (1992), Jain (1991), Lawler (1982), Rao (1997), Sood *et al.* (2006).

Dendrocalamus strictus **(Roxb.) Nees.**

Syn: *Bambusa stricta* Roxb.

Family: Poaceae

Eng., H. and Sans. Names : *Eng.*: Male bamboo, Solid bamboo.

H.: Bans, Banskaban, Banskhurd, Kopar, Narbans.

Sans.: Vansha, Venu, Yava-phala.

Distribution : Deciduous forests and in dry or moderately dry regions, practically all over India upto 1,200m.

Parts Used : Stem. Root.

Folk Uses : 7-21 pieces of its root tied with a sacred thread as a necklace for a week to cure typhoid. A piece of stem used as ear ring for magico-religious purposes. Strips of bamboo also tied on fractured bone of leg of animals to set it right.

Trade : Male bamboo.

References : Ambasta (1986), Anonymous (1952), Dhiman (2003), Jadhav (2008), Kirtikar and Basu (1984), Kumar *et al.* (2004), Patel *et al.* (2003), Sharma (2003).

Dendrophthoe falcata **(L.f.) Etting.**

Syn: *Loranthus falcatus* Linn. f.; *L. longiflorus* Desr.

Family: Loranthaceae

Eng., H. and Sans. Names : *Eng.*: Mistletoe.

H.: Banda.

Sans.: Vanda, Vrkshadani.

Distribution : Throughout India.

Parts Used : Stem. Plant.

Reproductive Cycle : December–April.

Folk Uses : Plant growing on 'Neem' tree collected during lunar or solar eclipse, and tied to the arm to prevent bad dreams. Likewise, the plant growing on a 'Bel' tree collected during solar eclipse and tied to the waist of the patient to relieve hysteria. Stem pieces worn as an amulet by 'adivasis' of Orissa to remove fever.

Active Constituents : Stem contains β-amyrin-OAc, oleanolic acid, its Me-ester acetate, β-sitosterol and stigmasterol.

References : Anonymous (1952), Asolkar *et al.* (1992), Chatterjee and Pakrashi (1997), Hembrom (1996), Jain (1991), Khan and Khanum (2005), Sen and Behera (2007), Sood *et al.* (2005).

Desmostachya bipinnata Stapf.

Syn: *Eragrostis cynosuroides* Beauv.

Family: Poaceae

Eng., H. and Sans. Names :	*Eng.*: Sacrificial grass. *H.*: Dab, Durva. *Sans.*: Darbha.
Distribution :	Known to occur in hot and dry places throughout India.
Part Used :	Grass.
Reproductive Cycle :	Flowering (August–November) and fruiting in cold season.
Folk Use :	'Kush' grass worn by couples at their wedding.
References :	Ambasta (1986), Mookerji (1930).

Dichrostachys cinerea (Linn.) Wt. and Arn.

Syn: *Acacia cinerea* Spreng.; *A. dalea* Desv.; *Cailliea cinerea* Macb.; *Desmanthus cinereus* Willd.; *Mimosa cinerea* Linn.

Family: Fabaceae

H. and Sans. Names :	*H.*: Bilatri, Kheri, Murmiar, Vurutuli. *Sans.*: Bahuvaraka, Dirghamula, Krichhari, Vellantara, Viravriksha.
Distribution :	Dry scrub forests, arid hills of N.W. and Peninsular India.
Part Used :	Stem bark.
Folk Use :	An amulet containing pieces of its stem bark and roots tied on right arm of children by Bhils and Garasias to protect from evil eyes.
Active Constituents :	Friedelin-α-amyrin and sitosterol isolated from bark.
References :	Anonymous 1952), Khan and Khanum (2005), Rastogi and Mehrotra (1970-1979), Singh and Pandey (1998), Sood and Sudershna (2008), Sood *et al.* (2005).

Didymosperma nanum H. Wdl. and Drude.

Syn: *Arenga nana* Griff.

Family: Arecaceae

Distribution :	Found in Arunachal Pradesh in N. E. India.
Parts Used :	Whole Plant.
Reproductive Cycle :	July–December.

Folk Use	:	Whole plant used for making women waist ring.
Reference	:	Ambasta (1986), Gangwar and Ramakrishnan (1990)

Diplocyclos palmatus **(Linn.) Jeff.**

Syn: *Bryonopsis laciniosa* (Linn.) Naud.; *Bryonia lanciniosa* Linn.; *B. palmata* Linn.

Family: Cucurbitaceae

Eng., H. and Sans. Names	:	*Eng.*: Bryony, Palmate diplocyclos. *H.*: Gargunaru, Sivavalli. *Sans.*: Bahupatra, Lingini.
Distribution	:	Throughout India on hedges and bushes upto 1,200m.
Parts Used	:	Root. Stem.
Reproductive Cycle	:	March–October
Folk Uses	:	Root collected during solar or lunar eclipse, tied to the waist of a woman for smooth delivery. Stem tied as necklace to cure fever by Khonds and Porjas. Necklace of roots worn as eye amulets for children.
References	:	Bennet (1986), Chatterjee and Pakrashi (1997), Modi (1893), Prajapati *et al.* (2006), Rama Rao and Henry (1996), Sen and Behera (2007).

Drypetes roxburghii **(Wall.) Hurus.**

Syn: *Putranjiva roxburghii* Wall.; *P. sphaerocarpa* Muell.

Family: Euphorbiaceae

Eng., H. and Sans. Names	:	*Eng.*: Life of the child tree. *H.*: Jiaputa, Jivputrak, Joti, Juti, Patigia, Patji, Putra-jiva. *Sans.*: Apatyajiva, Ardhasadhaka, Garbhada, Mavu, Pavitra, Putranjiva.
Distribution	:	Throughout tropical India.
Part Used	:	Fruits. Ripe Seeds.
Reproductive Cycle	:	Throughout year.
Folk Uses	:	A garland of moistened seeds hung around the neck to cure acidity and urticaria. Rosaries made from the stones of fruits put around the neck to preserve the wearer from any harm and to ward off diseases attributed to evil spirits and to cure skin allergy. Also necklace of fruits worn by a pregnant women to prevent miscarriage. Necklaces and rosaries also worn by Hindu saints and Brahmins.

Active Constituents : Saponin-putran-jivoside characterized from seed coat.

References : Bennet (1986), Bhatt *et al.* (2003), Chauhan (1999), Hembrom (1996), Jain (1991), Jha and Goel (2004), Khanna (2002), Mehra *et al.* (1975), Singh and Pandey (1998), Sinha and Sinha (2001), Sood *et al.* (2005).

Echinops echinatus **Roxb.**

Syn: *E. griffithianus* Boiss.

Family: Asteraceae

Eng., H. and Sans. Names : *Eng.*: Camel's thistle, Globe thistle.

H.: Gokhru, Utakanta, Utakatira.

Sans.: Brahma dandi, Kantalu, Kantaphala, Shrigala, Utati, Vrittaguchha.

Distribution : Throughout the plains of India.

Part Used : Root.

Reproductive Cycle : January–April.

Folk Uses : A small piece of root tied to the hair of the expectant mother at the time of delivery to induce labour pain and for normal delivery. For curing hoarse cough in children, root piece removed tied with a string around the neck of the children on Sunday. Likewise for curing typhoid, 7-21 pieces of its roots tied with a sacred thread around neck for a week. Uprooted plant root contacted over the hands of victim on scorpion sting. This plant is also used for tattooing.

Active Constituents : Root contains allophanic acid and methylallophanic acid.

References : Anonymous (1952), Bhatt *et al.* (2003), Jadhav (2008), Jain (1995), Jain *et al.* (1991), Jain and Sharma (1996), Khan and Khanum (2005), Punjani (2002), Singh and Pandey (1998), Sood and Sudershna (2008).

Eclipta prostrata **(Linn.) Hassk.**

Syn: *Cotula alba* Linn.; *E. alba* Hk.; *E. erecta* Lamk.; *E. marginata* Boin.; *Verbisina alba* Linn.; *V. prostrata* Linn.

Family: Asteraceae

Eng., H. and Sans. Names : *Eng.*: Trailing eclipta plant.

H.: Babri, Bengra, Bhangra, Moch-kand.

Sans.: Ajagara, Angaraka, Bhringraj, Kesharaja, Markara, Patanga, Shyamala.

Distribution	:	Common weed in moist situations throughout India.
Part Used	:	Leaves.
Reproductive Cycle	:	August–November.
Folk Use	:	Black dye from leaves used for tattooing.
Active Constituents	:	Important ingredients of leaves are α-terthienyl-methanol, stigmasterol, β-amyrin, and mixture of desmethyl wedelolactone.
Trade	:	Bhringraj.
References	:	Anonymous (1952), Barua *et al.* (2007), Bennet (1986), Chauhan (1999), Sharma (2003), Sood *et al.* (2005), Watt (1972).

Elaeagnus latifolia **Linn.**

Family: Elaeagnaceae

Eng. Name	:	*Eng.*: Bastard oleaster.
Part Used	:	Fruit.
Folk Use	:	Gold pendants made to imitate fruit.
References	:	Birwood (1880).

Elaeocarpus ganitrus **Roxb.**

Syn: *E. sphaericus* (Gaertn.) K. Schum.; *Ganitrus sphaericus* Gaertn.

Family: Elaeocarpaceae

Eng., H. and Sans. Names	:	*Eng.*: Utsarum bead tree.
		H.: Rudrak.
		Sans.: Bhutanashana, Harksha, Rudraksha.
Distribution	:	W. Ghats and Konkan region.
Parts Used	:	Nuts. Seeds.
Reproductive Cycle	:	April–December.
Folk Uses	:	An amulet of its seeds on neck, wrist, arm and chest, effective in controlling blood pressure, heart ailments, epilepsy and mental disabilities. Polished hard tubercled nuts made into rosaries, necklaces, buttons and bracelets and worn by Brahmins, holy men Hippies. Nuts worn by Shiva worshippers and also by widows. The only single lobed nut known worth a small fortune.
Active Constituents	:	Fruit reported to contain palmitic, isopalmitic, linoleic and myristic acid.

References : Anonymous (1952), Bhandarkar (1913), Bhatt *et al.* (2003), Bennet (1986), Bushan (1964), Chatterjee and Pakrashi (1997), Dhiman (2003), Jha and Goel (2004), Mehra *et al.* (1975), Sood *et al.* (2005), Vihari (1995).

Elaeocarpus lanceaefolius Roxb.

Family: Elaeocarpaceae

Distribution : E. Himalaya, hills of Assam upto 2,700m.

Part Used : Nuts.

Folk Use : Small nuts exported to Europe for beads, bracelets, necklaces.

References : Ambasta (1986), Anonymous (1952), Balfour (1871), Mehra *et al.* (1975).

Elaeocarpus tectorius (Lour.) Poir.

Syn: *Craspedum tectorium* Lour.; *E. leptostachya* Wall. *ex* C. Muell.; *E. robustus* Roxb.

Family: Elaeocarpaceae

Distribution : N. E. India upto 700m and in Andaman Islands.

Part Used : Seeds.

Folk Use : For making necklaces.

References : Anonymous (1952), Bennet (1986), Mittre (1997), Saklani and Jain (1994).

Elaeocarpus tuberculatus Roxb.

Family: Elaeocarpaceae

Distribution : Found from Kanara southwards through W. Ghats in Coorg, Mysore, Nilgiri, Palni and Anamalai hills.

Part Used : Nuts.

Folk Use : Nuts used for making necklaces, bracelets and beads.

References : Ambasta (1986), Anonymous (1952), Mehra *et al.* (1975).

Elephantopus scaber Linn.

Family: Asteraceae

Eng., H. and Sans. Names : *Eng.*: Prickly leaved elephant's foot.

H.: Gobhi, Gojilata, Karipadam.

Sans.: Gojihra, Kari padam, Kurasa, Satamulika.

Distribution : Hotter parts of India.

Part Used	:	Roots.
Reproductive Cycle	:	October–January.
Folk Uses	:	Pieces of root worn in ear by 'adivasis' of Orissa to cure headache. An entire root tied over forehead to get relief from headache and to the arm to prevent bad dreams. Root hung in the sleeping area of sucking children for better lactation in mother. An amulet of roots around neck also advised for children sickness.
Trade	:	Hastipadi. Gobhi.
References	:	Anonymous (1952), Augustine and Sivadasan (2004), Bhandary *et al.* (1996), Jain (1991), Khan and Khanum (2005), Maliya (2004), Sen and Behera (2007), Shah and Singh (1990), Singh and Prakash (1994), Sood *et al.* (2005).

***Elettaria cardamomum* Maton.**

Syn: *Alpinia cardamom* Roxb.

Family: Zingiberaceae

Eng., H. and Sans. Names	:	*Eng.*: Cardamom, Lesser cardamom. *H.*: Choti-elachi. *Sans.*: Ela, Upakunchika.
Distribution	:	Moist forest tracks upto 2,000m, commercially cultivated in Tamil Nadu.
Part Used	:	Seeds.
Reproductive Cycle	:	Almost throughout year.
Folk Use	:	Seeds strung into garlands for adornments in Thanjaver, Tamil Nadu.
Active Constituents	:	Seeds yield an essential oil, oleum cardamoni, with limonene, d-a-terpineol, borneol, cineol, and sabinene as important constituents.
References	:	Ambasta (1986), Francis (1984).

***Elytraria acaulis* (L.f.) Lindau.**

Syn: *E. crenata* Vahl.; *Tubiflora acaulis* Kuntze.; *E. lyrata* Vahl.; *Justica acaulis* L.f.

Family: Acanthaceae

H. Name	:	*H.*: Dasmori.
Distribution	:	Throughout India.
Part Used	:	Fresh roots.

Reproductive Cycle	:	July–March.
Folk Use	:	Fresh root collected during early morning or Sunday and passed seven times between the two legs of the medicineman and then tied to the cattle neck to hasten healing of decaying wounds.
References	:	Anonymous (1952), Bennet (1986), Jain *et al.* (1991), Kumar and Nagiyan (2006), Punjani (2002).

***Emilia sonchifolia* (Linn.) DC.**

Syn: *Cacalia sonchifolia* Linn.

Family: Asteraceae

Eng., H. and Sans. Names	:	*Eng.*: Red tassle flower, Sow thistle flower. *H.*: Hirankhur, Hirankhuri. *Sans.*: Sasasruti, Sasasrutih.
Distribution	:	Cultivated fields and open habitats throughout most of India upto 1,200m.
Part Used	:	Whole Plant.
Reproductive Cycle	:	Flowering and fruiting in rainy season
Folk Use	:	An amulet of the whole plant around neck cures fever in children below one year.
Active Constituents	:	Reported constituents from the plant are kaempferol-3-β-D-galactoside, quercitrin, rutin, quercetin, ursolic acid, n–hexacosanol and triacontane.
References	:	Anonymous (1952), Hembrom (1996), Khan and Khanum (2005), Rastogi and Mehrotra (1980-1984), Varghese (1996).

***Enicostemma axillare* (Lam.) Raynal.**

Syn: *Adenema hyssopifolium* G.Don.; *E. hyssopifolium* (Willd.) I.C Verdoorn.; *E. littorale* auct. non Bl.; *Exacum hyssopifolium* Willd.; *Gnetiana verticillata* Linn.

Family: Gentianaceae

Eng., H. and Sans. Names	:	*Eng.*: Indian gentian. *H.*: Chota cirayita, Katvinayi, Nai. *Sans.*: Mamajjakah, Magajihva, Nagajihva, Nahi, Tiktapatra.
Distribution	:	Throughout India from Panjab to Kanyakumari.
Part Used	:	Leave.

Reproductive Cycle	:	Fruiting throughout the year.
Folk Use	:	Leaves tied on the arm as an amulet by 'Bhils' and 'Garasia' tribals of Rajasthan to cure fever.
References	:	Bennet (1986), Khare (2007), Prajapati *et al.* (2006), Singh and Pandey (1998), Sood *et al.* (2005).

Ensete ventricosum (Welw.) Cheesman.

Syn: *E. edule* Horan.; *Musa ensete* J.F. Gmelin.

Family: Musaceae

Eng. Name	:	*Eng.*: False banana.
Distribution	:	N. East India at high altitudes.
Part Used	:	Seeds.
Folk Use	:	Seeds used for making necklaces.
References	:	Ambasta (1986), Ramachandran *et al.* (1986).

Entada phaseoloides (Linn.) Merr.

Syn: *E. scandens* Benth.; *E. pursaetha* DC.; *Mimosa scandens* (Linn.) Roxb

Family: Fabaceae

Eng., H. and Sans. Names	:	*Eng.*: Climbing entada, Elephant creeper, Garbee bean, Mackay bean, Nicker bean. *H.*: Chian, Garbi, Gila. *Sans.*: Bidhanta.
Distribution	:	Central and E. India, Sikkim, Western Peninsula.
Part Used	:	Seeds.
Reproductive Cycle	:	March–May onwards.
Folk Use	:	Ornaments made from its seeds worn as charm by tribe people.
Active Constituents	:	Entagenic acid characterized from kernels.
References	:	Ambasta (1986), Mehra *et al.* (1975), Sood *et al.* (2009).

Equisetum debile Roxb. *ex Vauch.*

Syn: *E. ramossisimum* Desf.

Family: Equisetaceae

Eng. Name	:	*Eng.*: Many branched horsetail.
Distribution	:	Throughout Himalaya.
Part Used	:	Stems.

Folk Use : Tribals believe that the broken bones if tied around by the stems of this plant, get joined up.

Trade : Fox tailed asparagus.

References : Anonymous (1952), Jain and Mudgal (1999), Rastogi and Mehrotra (1980-1984), Sharma (2003).

Ervatamia heyneana **(Wall.) Cooke.**

Syn: *Tabernaemontana alternifolia* Linn.; *T. heyneana* Wall.

Family: Apocynaceae

H. Name : *H.*: Nagkuda.

Distribution : Konkan, W. Ghats in Malabar.

Parts Used : Fruit. Flowers.

Reproductive Cycle : March–December.

Folk Uses : Fruits tied with thread and put around the neck in children by Girijanas to cure cough. Flowers worn in the hair.

Active Constituents : Fruits exhibited lupeol-OAc and β-Amyrin-OAc.

References : Anonymous (1952), Asolkar *et al.* (1992), Francis (1984), Kshirsagar and Singh (2000, 2007).

Erythrina indica **Lam.**

Syn: *E. corallodendron* Linn.; *E. cuneata* Grah.; *E. mysorensis* Gamble.; *E. spathacea* Wall.; *E. stricta* Roxb.; *Micropteryx stricta* Walp.; *E. orientalis* Linn.; *E. variegata* Linn.; *E. variegata* var. *alba* (Blatt. and Mill.) Mahes.; *E. variegata* var. *marmorata* (Vietch.) Mahes.; *E. variegate* var. *parcellii* (Hort.) Mahes

Family: Fabaceae

Eng., H. and Sans. Names : *Eng.*: Indian coral tree, Mochi wood.

H.: Dadap, Ferrud, Mandara, Pangara, Pharad.

Sans.: Bahupushpa, Kantaki, Krimighna, Palitmandar, Parijata.

Distribution : Coastal forests from Bombay to Malabar, Sunderbans, Andaman and Nicobar Islands.

Parts Used : Flowers. Beads. Wood. Bark.

Reproductive Cycle : March–April.

Folk Uses : Pieces of bark tied around the neck of the newly born child to ward off evil spirits in Manipur. Beads employed for making necklaces. Onges place flowers on their head for adornments. A piece of wood tied on the neck of animal to cure wounds.

Active Constituents : These are erysovine, erysodine, hypaphorine, n-alkanes, n-alkanols, alkyl-ferulates, fattyacids, esters, sterols and a coumarin (bark), hypaphorine, erysopine and erysotrine (seeds).

Trade : Coral tree.

References : Anonymous (1952), Asolkar *et al.* (1992), Awasthi and Goel (2000), Bennet (1986), Jain (1991), Kirtikar and Basu (1984), Mehra *et al.* (1975), Mishra *et al.* (1996), Rastogi and Mehrotra (1970-1979), Sharma (2003), Sood *et al.* (2005).

Erythrina suberosa Roxb.

Syn: *E. alba* Roxb.; *E. glabrescens* Prain.; *E. nahasuta* Ham.; *E. reniformis* Ham.; *Micropteryx suberosa* Walp.

Family: Fabaceae

Eng., H. and Sans. Names : *Eng.*: Indian coral tree.

H.: Dauldhak, Mandara, Nasut, Rowanra, Rungra.

Sans.: Bahupushpa, Kantaki, Parijata.

Distribution : Throughout India.

Part Used : Flowers.

Reproductive Cycle : March–June.

Folk Use : Flowers considered good for adornments among tribals in the form of garlands along the Aravalli range.

Active Constituents : Flowers found to contain delphinidin-3,5-diglucoside, cyanin-3,5-digluco-side, erysodine, erysopine, erythraline, and erthratidine.

References : Anonymous (1952), Bennet (1986), Rastogi and Mehrotra (1970-1994), Singh and Pandey (1998), Sood *et al.* (2009).

Eucalyptus maculata Hk.

Syn: *E. citriodora* Hk.

Family: Myrtaceae

Eng. Name : *Eng.*: Lemon scented eucalypt.

Distribution : Cultivated in many parts of India.

Part Used : Leaf.

Folk Use : Leaves adorned for hair decoration.

Active Constituents : Leaves contain eucalyptin, β-sitosterol, betulinic and ursolic acids, and citriodorin.

References : Anonymous (1952), Asolkar *et al.* (1992), Rastogi and Mehrotra (1970-1979, 1980-1984), Saklani and Jain (1994), Sharma (2003).

Euclea fruiticosa Forssk.

Family: Ebenaceae

Part Used : Twigs.

Folk Use : Twigs used to make lips and mouth very red.

Reference : Rajan *et al.* (2005).

Eugenia caryophyllata Thunb.

Syn: *Caryophyllus aromaticus* Linn.; *Syzygium aromaticum* (Linn.) Merr.

Family: Myrtaceae

Eng. and H. Names : *Eng.*: Clove tree.

H.: Laung.

Part Used : Flowers.

Folk Use : Flower buds strung into garlands in Thanjaver, Tamil Nadu.

References : Ambasta (1986), Francis (1984).

Euonymus grandiflorum Wall.

Syn: *E. lacerus* Ham.

Family: Celastraceae

Distribution : Found in Himalaya and Assam.

Part Used : Seeds.

Folk Use : Seeds strung as beads for adornment.

References : Anonymous (1952), Bennet (1986), Jain (1991), Watt (1889, 1972).

Euonymus fimbriatus Wall.

Family: Celastraceae.

Part Used : Seeds.

Folk Use : Seeds worn as beads.

Reference : Watt (1889).

Euphorbia ligularia Roxb.

Syn: *E. nerifolia* Linn.

Family: Euphorbiaceae

Eng., H. and Sans. Names : *Eng.*: Holy milk hedge.

H.: Pattonisend, Sehund, Thohar, Thuar.

Sans.: Gudaa, Sehunda, Snuhi, Sudhaa, Vajjri, Vajra.

Distribution : Wild in C. India, Orissa and S. India; cultivated in W. Bengal.

Parts Used : Root. Stem.

Reproductive Cycle : During Spring.

Folk Uses : A small piece of root inserted into the vagina of a pregnant woman for inducing abortion. Tribals of M.P. use its stem for whitlow ('khurni') by making a hollow cup out of its stem and the affected finger inserting in it for a few days.

Active Constituents : Properties of root due to cycloartenol, euphorbol, ingenol triacetate, 12-deoxyphorbol-13,20-di acetate, delphinidin-3,5-diglucoside, 24-methylenecycloartenol, tulipanin-3,5-diglucoside whereas stem contains new triterpene-glut-5(10)-en-1-one, friedelan-3α and 3β-ol and taraxerol.

References : Barua *et al.* (2000), Bennet (1986), Jain *et al.* (1991), Rastogi and Mehrotra (1970-1984), Sen and Behera (2007), Sood and Sudershna (2008).

Euphorbia prostrarta Ait.

Syn: *E. chamaesyce* Linn.

Family: Euphorbiaceae

Part Used : Stem latex.

Folk Use : Used for tattooing.

References : Bennet (1986), Jain (1995), Jain and Sharma (1996).

Evolvulus nummularius (Linn.) Linn.

Syn: *Volvulopsis nummularis* (Linn.) Roberty.

Family: Convolvulaceae

Distribution : Throughout plains of India and Nepal.

Part Used : Whole Plant.

Reproductive Cycle	:	August–June.
Folk Use	:	Whole plant tied around the head to get sound sleep.
Active Constituents	:	Plants rich in new flavone–7,3′,4′,5′-tetrahydroxy-flavone and its 2 glycosides-evolvusides A and B.
References	:	Rastogi and Mehrotra (1985-1989), Sen and Behera (2007), Varghese (1996).

Ficus altissima **Blume.**

Syn: *F. laccifera* Roxb.; *Urostigma altissimum* Miq.; *U. lacciferum* Miq.

Family: Moraceae

Eng. Name	:	*Eng.*: Council tree.
Distribution	:	Tropical Himalaya, plains and lower hills of Deccan Peninsula, Assam, Mayanmar and Andaman Islands.
Part Used	:	Bark.
Reproductive Cycle	:	November–May.
Folk Use	:	Bark of aerial root used for making amulets. Bark fibers also used to string human bone necklaces in the Andaman Islands.
References	:	Radcliffe-Brown (1922), Sood *et al.* (2009).

Ficus hispida **Linn.**

Syn: *F. daemonum* Koenig.; *F. mollis* Willd.; *F. oppositifolia* Willd.; *Covellia assamica* Miq.; *C. courtallensis* Miq.; *C. oppositifolia* Gasp.

Family: Moraceae

Eng., H. and Sans. Names	:	*Eng.*: Devil fig. *H.*: Daduri, Dagurin, Gobla, Kagsha, Katgularia. *Sans.*: Kakadumbura, Ummialto-dumbara.
Distribution	:	Throughout India.
Part Used	:	Bark.
Folk Use	:	Bark tassels used for adornments for waist, neck, and arms in Little Andamans.
Active Constituents	:	Bark contains tannin and fibres.
References	:	Ambasta (1986), Radcliffe-Brown (1922), Sood *et al.* (2005).

Ficus racemosa Linn.

Syn: *Covellia glomerata* Miq.; *C. mollis* Miq.; *F. chitagonga* Miq.; *F. glomerata* Roxb.; *F. goolereca* Roxb.; *F. mollis* Miq.

Family: Moraceae

Eng., H. and Sans. Names	:	*Eng.*: Cluster fig, Gular fig. *H.*: Dimere, Gular, Lelka, Paroa, Umarai. *Sans.*: Hemdugdhak, Jantuphala, Phalasambendha, Udumbara.
Distribution	:	Throughout India.
Parts Used	:	Leaves. Bark. Root. Seeds.
Reproductive Cycle	:	April–July.
Folk Uses	:	Leaves used for adornments. Bark and leaves chewed for colouring lips. Root collected on the early morning of a Saturday from the east of the plant and tied on the ear with a thread opposite to the affected parts to heal dental carries. Women wear seeds on eighth day of confinement.
Active Constituents	:	Properties due to ceryl behenate, lupeol, α-amyrin and 3 unidentified compounds (bark), gluanol, β-amyrin and β-sitosterol (leaves).
References	:	Asolkar *et al.* (1992), Auboyer (1961), Chattopadhyaya (1976), Gogoi and Borthakur (1991), Jain *et al.* (1991), Saklani and Jain (1994), Sen and Behera (2007), Sood *et al.* (2009), Watt (1889).

Ficus religiosa Linn.

Syn: *F. affinior* Griff.; *Urostigma affine* Miq.; *U. religiosum* Gasp.

Family: Moraceae.

Eng., H. and Sans. Names	:	*Eng.*: Bot tree, Pipul tree, Sacred bo tree, Sacred fig, The peepul tree. *H.*: Pipal, Piple. *Sans.*: Ashwattha, Chaityavriksha, Pavitraka, Pippal, Yajnika.
Distribution	:	Throughout India upto 1,400m.
Part Used	:	Leaves.
Reproductive Cycle	:	March–October.
Folk Uses	:	Young leaves used in snake bite; it is believed that the leaf petiole when inserted in both the ears sucks poison

		from the body of victim. 'Garasias' place its leaves on the head of lady in labour pains to facilitate the delivery. Harappan 'winged beads' perhaps modeled on the leaves. Leaves reported as only clothing of aboriginal women.
References	:	Bhatt *et al.* (2003), Birwood (1880), Dikshit (1952b), Jadhav (2008), Jain *et al.* (1991), Sikarwar and Kaushik (1992), Singh and Pandey (1998), Sood *et al.* (2005, 2009).

Flacourtia ramontchi Linn.

Syn: *F. indica* Merr.; *F. sepiara* Roxb.; *F. latifolia* Cooke.; *F. occidentalis* Blatter.

Family: Flacourtiaceae

Eng., H. and Sans. Names	:	*Eng.*: Governor's Plum, Madagascar Plum.
		H.: Bilangra, Kanju.
		Sans.: Aghori.
Distribution	:	Scrub forests and rocky hills upto 900m.
Part Used	:	Wood.
Reproductive Cycle	:	March–April.
Folk Uses	:	Wood cut into beads for rosaries and necklaces.
References	:	Ambasta (1986), Sood *et al.* (2009), Watt (1889).

Flemingia bracteata (Roxb.) Wt.

Syn: *F. fruticulosa* Wall. *ex* Benth.; *F. strobilifera* R.Br. var. *bracteata* Baker.; *Hedysarum strobiliferum* Linn.; *Mogania bracteata* Hui-Linn.; *M. strobilifera* (Linn.) St. Hill *ex* Jacker.

Family: Fabaceae

H. Name	:	*H.*: Kusrunt.
Distribution	:	Himalaya (Shimla–Kumaon) ascending to 2,700m. to Assam, Khasia.
Part Used	:	Root.
Reproductive Cycle	:	January–November.
Folk Use	:	Roots kept under the bed of pregnant lady for easy delivery.
References	:	Ambasta (1986), Anonymous (1956), Bennet (1986), Hooker (1897), Sood and Sudershna (2008), Varghese (1996).

Flemingia chappar **Buch. Ham. *ex* Benth.**

Syn: *Moghania chappar* Buch. Ham.

Family: Fabaceae

H. Name	:	*H.*: Salpan.
Distribution	:	E. Himalaya, Bihar.
Part Used	:	Twig.
Reproductive Cycle	:	January–March onwards.
Folk Use	:	A piece of twig kept underneath the pillow for sound sleep.
Trade	:	Salpan.
References	:	Jain *et al.* (1991), Jha and Varma (1996), Sahoo and Bahali (2003).

Flueggea leucopyrus **Willd.**

Syn: *Securinega leucopyrus* (Willd.) Muell.–Arg.

Family: Euphorbiaceae

Eng., H. and Sans. Names	:	*Eng.*: Spinous fluggea. *H.*: Ainta, Hartho. *Sans.*: Svetakamboja.
Distribution	:	Throughout India.
Part Used	:	Seed.
Folk Use	:	Chenchus wear the seeds of this plant.
References	:	Ambasta (1986), Furer-Haimendorf (1945), Prajapati *et al.* (2006).

Garcinia morella **Dear.**

Syn: *G. acuminate* Planch and Train.; *G. elliptica* Wall.; *G. lobulasa* Wall.; *G. pictorial* Roxb.

Family: Clusiaceae

Eng., H. and Sans. Names	:	*Eng.*: Indian Gamboge Tree, True Gambose. *H.*: Ghotajhauba, Gota ganba, Tamal. *Sans.*: Amritadruma, Kankustha, Tamala.
Distribution	:	Bengal, Khasia Hills and both the Peninsulas.
Part Used	:	Flowers.
Reproductive Cycle	:	December–June.
Folk Use	:	Flowers used for adornments.
References	:	Ambasta (1986), Dymock (1889a), Sood *et al.* (2009).

Gardenia turgida Roxb.

Family: Rubiaceae.

H. and Sans. Names	:	*H.*: Thanela. *Sans.*: Bharangi.
Distribution	:	Throughout greater parts of India in dry open deciduous forests, ascending to 1,330m in Outer Himalaya.
Part Used	:	Roots.
Reproductive Cycle	:	April–June.
Folk Use	:	Roots of this tree strung with a cord and worn as charm by tribal people of Chotta Nagpur.
Active Constituents	:	D–mannitol, its hexaacetate, hexabenzoate, gardenin A,B and E, oleanolic acid, α-amyrin, β-sitosterol isolated from roots.
References	:	Anonymous (1956), Khan and Khanum (2005), Mehra *et al.* (1975).

Geranium himalayense Klot.

Syn: *G. palustre* auct. non Linn.; *G. pratense* auct. non Linn.

Family: Geraniaceae

Eng. and H. Names	:	*Eng.*: Cranes bill, Crow foot, Grace of god, Gratia dei, Meadow cranes bill. *H.*: Bhanda.
Distribution	:	W. Temperate Himalaya.
Part Used	:	Flowers.
Folk Use	:	As adornments.
References	:	Bennet (1986), Jain (1991), Kirtikar and Basu (1984).

Globba marantina Linn.

Syn: *G. bulbifera* Roxb.

Family: Zingiberaceae

Distribution	:	Khasia hills, Meghalaya.
Part Used	:	Flowers.
Reproductive Cycle	:	August.
Folk Use	:	Flowers worn by women.
References	:	Bennet (1986), Chandra (1997), Hosagoudar and Henry (1996a).

Gloriosa superba **Linn.**

Syn: *Eugona superba* Salisb.; *G. abyssinica* A. rich.; *G. doniana* Schultes.; *G. homblei* Willd.; *Methonica doniana* Kunth.; *M. superba* Lamk.

Family: Liliaceae

Eng., H. and Sans. Names	:	*Eng.*: Climbing lily, Malabar glory lilly, Superb lily, Tiger's claw. *H.*: Kariari, Karidhari, Kulhari, Languli, Ravigani. *Sans.*: Agniskha, Agnimukhi, Ailni, Amrita, Halini, Indrapushpi, Langalika.
Distribution	:	N. W. Himalaya–Assam.
Parts Used	:	Flowers. Root. Stem. Tubers.
Reproductive Cycle	:	September–October.
Folk Uses	:	Onge women adorn their hairdo with flowers of this plant. Root pieces of about 5 cm tied to the waist of women to facilitate delivery and removed immediately after delivery. Pieces of its stem tied to the arm and its tuber kept below the abdomen of women in labour to facilitate the child birth. Garland of fresh tuber pieces around neck of patient believed to cure jaundice.
Active Constituents	:	Flowers, roots and tubers contain colchicine.
Trade	:	Glory lily.
References	:	Bhargava (1997), Chakrabarty and Balakrishnan (2003), Chaudhari and Hutke (2002), Chauhan (1999), Jain (1991), Jain and Singh (1997), Kshirsagar and Singh (2007), Ranjan (1996), Singh and Pandey (1998), Sinha and Sinha (2001), Sood and Sudershna (2008), Sood *et al.* (2005).

Gnetum scandens **Roxb.**

Syn: *G. edule* Blume.

Family: Gnetaceae

Part Used	:	Fiber.
Folk Use	:	Fibers used as amulet on the Andaman Islands.
Reference	:	Radcliffe-Brown (1922).

***Gnidia glauca* (Fresen.) Gilg.**

Syn: *G. eriocephalus* Meis.; *G. insularis* Gardn.; *Lasiosiphon eriocephalus* (Meissn.) Decne.; *L. glaucus* Fresen.

Family: Thymelaeaceae

Distribution	:	W. Ghats.
Part Used	:	Stem.
Folk Use	:	Stem tied around the stomach to cure dropsy by Kaniyanas.
Active Constituents	:	Stem rich in lasiocephalin.
References	:	Bennet (1986), Khare (2007), Kshirsagar and Singh (2007), Rastogi and Mehrotra (1970-1979).

***Gossypium arboreum* Linn.**

Syn: *G. nanking* Meyen.; *G. indicum* Tod.; *G. neglectum* Tod.; *G. sanguineum* Hassk.

Family: Malvaceae

Eng., H. and Sans. Names	:	*Eng.*: Tree cotton.
		H.: Kapas, Rui, Tula.
		Sans.: Karpasamu.
Distribution	:	Grown in gardens and temples.
Part Used	:	Fruits.
Folk Uses	:	Fibres obtained from the fruits of this plant used for hair and ears decoration by Nagas. Cotton fibers plaited into collars by Angami Naga girls to give to their boy friends Also fibers spun into the sacred thread and worn on the neck as amulets throughout India.
References	:	Ambasta (1986), Buchanan (1899), Hodson (1911), Hutton (1921a and b), Mehra *et al.* (1975), Mills (1926).

***Grewia subinaequalis* DC.**

Syn: *G. asiatica* Linn.; *G. elastica* Royle.; *G. hainesiana* Hole; *G. tiliaefolia* Vahl.

Family: Tiliaceae

Eng., H. and Sans. Names	:	*Eng.*: Asiatic grewia.
		H.: Dhamin, Karsa, Parusha, Phalsa, Pharsa.
		Sans.: Alpasthi, Dhanvanchhada, Giripilu, Mriduphala.
Distribution	:	Deciduous forests in the sub-Himalayan region and Peninsular India.

Parts Used	:	Root. Stem.
Reproductive Cycle	:	March–April.
Folk Use	:	A small piece of its stem or root worn as an amulet on the waist for curing inflammation of the testicles.
Active Constituents	:	Active principles of stem attributed to lupeol, lupenone, betulin and friedelin.
Trade	:	Phalsa.
References	:	Anonymous (1956), Bennet (1986), Chatterjee and Pakrashi (1997), Kirtikar and Basu (1984), Rastogi and Mehrotra (1970-1979), Tarafder *et al.* (1997).

Guazuma ulmifolia Lamk.

Syn: *G. tomentosa* Kunth.; *Theobroma guazuma* Linn.

Family: Sterculiaceae

Eng. Name	:	*Eng.*: Bastard cedar.
Distribution	:	Cultivated as a roadside shade tree in warmer parts of country.
Part Used	:	Fruit.
Folk Use	:	Fruits worn by the Soliga men and women as beads in the form of garlands or as an ornament.
References	:	Bennet (1986), Gopal and Jagdishchandra (2003).

Guettarda speciosa Linn.

Syn: *Cadamba jasminiflora* Sonner; *Jasminum hirsutum* Willd.; *Nictanthes hirsuta* Linn.

Family: Rubiaceae

Eng., H. and Sans. Names	:	*Eng.*: Beautiful guettarda, Velvet seed.
		H.: Domdomah.
		Sans.: Himma.
Distribution	:	Along the shores of S. India, Andaman Islands.
Part Used	:	Flowers.
Folk Use	:	Onge women adorn their hairdo with the flowers.
References	:	Chakrabarty and Balarkishnan (2003), Dhiman (2003).

Gyrocarpus americanus Jacq.

Syn: *G. asiaticus* Willd.; *G. jacquinii* Gaertn

Family: Combretaceae

H. Name	:	*H.*: Zaitun.
Distribution	:	Deccan peninsula, Orissa, Bengal, Andamans.

Part Used : Seeds.

Folk Use : Necklaces and rosaries formed by seeds in various parts of India.

References : Ambasta (1986), Bennet (1986), Mehra *et al.* (1975), Watt (1889.

***Haldina cordifolia* (Roxb.) Ridsdale.**

Syn: *Adina cordifolia* (Roxb.) Hk. f. *ex* Brandis.; *Nauclea cordifolia* Roxb.

Family: Rubiaceae

Eng., H. and Sans. Names : *Eng.*: Haldu.

H.: Haldu, Kadami.

Sans.: Dharakadamba, Haridaru.

Distribution : Himalaya to Central and S. India; common in E. Ghats, Mysore, Parts of Konkan.

Part Used : Leaves.

Reproductive Cycle : June–March.

Folk Use : An amulet of its leaves worn on wrist as a band in Birhor friendship ceremony.

References : Bennet (196, Dhiman (2003), Prajapati *et al.* (2006), Roy (1925), Sood *et al.* (2005), Varghese (1996).

***Helianthus annuus* Linn.**

Family: Asteraceae

Eng., H. and Sans. Names : *Eng.*: Lady eleven o'clock, Marigold of Peru, Sunflower.

H.: Hurhuja, Surajmukhi.

Sans.: Adityabhakta, Suryamukhi, Suryavarta, Suvarchala.

Distribution : Throughout India.

Part Used : Flowers.

Reproductive Cycle : April–August.

Folk Use : Used as adornments.

Active Constituents : Properties of flowers due to diterp. acid, trachyloban-19-oic acid, 3 new bisdesmosidic triterpenoid saponins–helianthosides 1,2 and 3.

References : Ambasta (1986), Anonymous (1994), Asolkar *et al.* (1992), Gogoi and Borthakur (1991), Kirtikar and Basu (1984), Rastogi and Mehrotra (1970-1994).

Helicteres isora **Linn.**

Syn: *H. chrysocalyx* Miq.; *H. roxburghii* G. Don.

Family: Sterculiaceae

Eng., H. and Sans. Names	:	*Eng.*: East India screw tree. *H.*: Avarttani, Bhendu, Javitri, Kapasi, Merosi. *Sans.*: Aavartaki, Aavartphala, Avartanee, Mrigashinga.
Distribution	:	Throughout India.
Part Used	:	Fruit.
Reproductive Cycle	:	May–January.
Folk Uses	:	Fruit tied as an amulet in neck to treat disease of malnutrition among children (locally known as 'dubli' disease), relieving post delivery pain and for curing scabies. Also, its garland tied to the waist of children to check bleeding from nose. An amulet of its 4 fruits tied to each hand and foot of children below 6 months age for curing polio, and around neck of children suffering from giddiness and tenesmus in stomach.
Active Constituents	:	Active properties of fruit attributed to diosgenin, α and β-Amyrins, friedelin, lupeol and taraxerone.
Trade	:	East Indian screw tree.
References	:	Anonymous (1959), Chauhan (1999), Dhiman (2003), Girach and Aminuddin (1995), Hembrom (1996), Hosagoudar and Henry (1996a), Kshirsagar and Singh (2007), Sen and Behera (2007), Sharma (2003), Sinha and Sinha (2001).

Helminthostachys zeylanica **(Linn.) Hk.**

Syn: *H. dulcis* Kaulf.; *Osmunda zeylanica* Linn.

Family: Ophioglossaceae

Distribution	:	Widely distributed in India.
Part Used	:	Root.
Folk Use	:	Root tied to the neck for swelling of the joints due to rheumatism.
Active Constituents	:	Main ingredients of roots are stigmasterol, fucosterol and dulcitol.
References	:	Ambasta (1986), Jain *et al.* (1991), Molla and Roy (1996), Prajapati *et al.* (2006), Rastogi and Mehrotra (1980-1984).

Hemidesmus indicus **(Linn.) Schult.**

Syn: *Asclepias pseudo-sarsa* Roxb.; *H. wallichii* Miquel.; *Periploca indica* Willd.

Family: Asclepiadaceae

Eng., H. and Sans. Names : *Eng.*: False sarsaparilla, Indian sarsaparilla.

H.: Annantmul, Hindi salsa, Magrabu, Salsa, Shwet-sariva.

Sans.: Ananta, Gopi-mulam, Naga-jihva, Sariva, Sugandhi.

Distribution : Upper Gangetic Plains, Eastwards to W. Bengal.

Parts Used : Root. Whole Plant.

Reproductive Cycle : July–December.

Folk Uses : An amulet of small pieces of root around the neck checks fever and ward off the evil spirits in Bastar district. Whole plant kept below the head while sleeping for inducing lactation by Soliga women.

Active Constituents : Properties of roots due to hexatriacontane, lupeol, its octacosanoate, α-amyrin, β-amyrin, its acetate and sitosterol

Trade : Country sarsaparilla.

References : Anonymous (1959), Jain (1963, 1991), Kirtikar and Basu (1984), Kshirsagar and Singh (2007), Rastogi and Mehrotra (1970-1979), Sen and Behera (2007), Sinha and Sinha (2001), Sood *et al.* (2005).

Hevea brasiliensis **Muell.–Arg.**

Family: Euphorbiaceae

Eng. Names : *Eng.*: Caoutchouc Tree, Para Rubber Tree.

Part Used : Rubber.

Folk Use : Nagas wear bangles of hardened rubber.

References : Ambasta (1986), Furer-Haimendorf (1933).

Hibiscus rosa-sinensis **Linn.**

Syn: *H. arnotii* Griff.; *H. liliiflorus* Griff.

Family: Malvaceae

Eng., H. and Sans. Names : *Eng.*: Chinese Hibiscus, Shoe flower, Scarlet rose-mallow.

H.: Gulhar, Jasum, Jasut, Lal ambari.

		Sans.: Hariballabha, Odhrapushpa, Pratika, Rudrapushpa, Trisandhya.
Distribution	:	Cultivated throughout India.
Part Used	:	Flowers.
Reproductive Cycle	:	February–October.
Folk Use	:	Used as adornments.
Active Constituents	:	Flowers rich in quercetin-3-diglucoside 3,7-diglucoside, cyaniding-3,5-diglucoside and cyanin-3-sophoroside-5-glucoside (deep yellow flowers) and kaempferol-3-xylosylglucoside (ivory white flowers).
Trade	:	China rose.
References	:	Ambasta (1986), Anonymous (1959), Gogoi and Borthakur (1991), Rastogi and Mehrotra (1970-1979), Sharma (2003), Sood *et al.* (2009).

Hibiscus tiliaceus **Linn.**

Syn: *H. tortuosus* Roxb.; *Paritium tiliaceum* St. Hill.

Family: Malvaceae

Eng., H. and Sans. Names	:	*Eng.*: Coast Cotton Tree, Yellow Mallow Tree.
		H.: Bola, Chelwa.
		Sans.: Japa.
Distribution	:	Tropical regions especially near the coast.
Part Used	:	Fiber.
Folk Use	:	Fibers used for men's belts in the Andaman Islands.
References	:	Ambasta (1986), Kirtikar and Basu (1984), Radcliffe-Brown(1922).

Hiptage benghalensis **(Linn.) Kurz.**

Syn: *Banisteria benghalensis* Linn.; *Gaertnera racemosa* Roxb.; *H. madablota* Gaertn.; *H. parvifolia* W. and A.

Family: Malpighiaceae

Eng., H. and Sans. Names	:	*Eng.*: Clustered hiptage.
		H.: Atimukta, Kampti, Madholota, Madmatti.
		Sans.: Kamuka, Madahvi, Malati, Mandapa.
Distribution	:	Assam, Chennai, E. Bengal, Konkan, Kumaon, Mumbai, Karnataka, Andamans.
Part Used	:	Flowers.

Reproductive Cycle	:	February–May.
Folk Use	:	Flowers employed for hair decoration.
References	:	Ambasta (1986), Anonymous (1959), Bennet (1986), Dhiman (2003), Saklani and Jain (1994).

Holarrhena antidysenterica Wall. *ex* DC.

Syn: *Echites antidysentrica* Roxb.; *E. pubescens* Buch-Ham.; *H. malaccensis* Wight.; *H. pubescens* Wall. *ex* Don; *Wrightia antidysentrica* Grah.

Family: Apocynaceae

Eng., H. and Sans. Names	:	*Eng.*: Antidysentric holarrhena, Conessi, Conessi bark, Kurchi tree. *H.*: Dhudi, Hat, Indrajau, Karchi, Kaura, Kura. *Sans.*: Girimallika, Kutaja, Mallikapushpa, Shakra-paryaya.
Distribution	:	Throughout plains of India.
Parts Used	:	Flowers. Leaves. Root.
Reproductive Cycle	:	May–October.
Folk Uses	:	A garland of flowers used as hairdo adornments both by men and women and also worn on ears. Paste of leaves used for colouring palms alongwith leaves of *Ehretia laevis.* A piece of root placed in the hair to ease delivery.
Active Constituents	:	These are kurchiphylline, kurchiphyllamine, holadysine and holadysamine (leaves), and holacetine (root).
Trade	:	Indrajau karwaa.
References	:	Anonymous (1959), Dhiman (2003), Girach and Aminuddin (1995), Sharma (2000), Singh and Pandey (1998), Sinha and Sinha (2001), Sood *et al.* (2009).

Holoptelea integrifolia (Roxb.) Planch.

Syn: *Ulmus integrifolia* Roxb.

Family: Ulmaceae

Eng., H. and Sans. Names	:	*Eng.*: The entire leaved Elm, The Indian Elm. *H.*: Banchilla, Begana, Chilbil, Chilmil, Karanji, Kumba, Kunj, Papri. *Sans.*: Chirabilva.

Distribution : Throughout India, sub-Himalaya, Ajmer, Bundelkhand, Bihar, Assam.

Parts Used : Bark. Stem.

Folk Use : A piece of stem wood of about 2 cm length hung around the neck of women as a charm to cure them of madness. Bark tied symbolically to arm to cure hydrocele.

Active Constituents : Bark characterized by the occurrence of friedelin, triterpenoid, fatty acid esters-holoptelin A and B, β-sitosterol.

References : Jain *et al.* (1991), Khan and Khanum (2005), Rastogi and Mehrotra (1970-1979), Saxena *et al.* (1997), Singh and Narain (2008), Sood *et al.* (2005).

Hordeum vulgare **Linn.**

Syn: *H. sativum* Jessen.; *H. hexastichon* Linn.; *H. intermedium* Kornicke; *H. distichon* Linn.

Family: Poaceae

Eng., H. and Sans. Names : *Eng.*: Barley.

H.: Jao, Jau, Jav, Jawa.

Sans.: Situshuka,Yava, Yavaka.

Distribution : Cultivated mainly in N. India and upto 4,350m in the Himalaya.

Part Used : Whole Plant.

Folk Use : Birhor girls put plant behind each other's in friendship ceremony.

References : Ambasta (1986), Roy (1925), Watt (1889).

Hornstedtia arunachalensis **Sunil and Ved Prakash.**

Family: Zingiberaceae

Part Used : Flowers.

Folk Use : Flowers used during marriage ceremony for adornment as ornamentals by Adi tribal women.

Reference : Tripathi and Goel (2001).

Hydrocotyle javanica **Thunb.**

Syn: *H. himalaica* Muker.; *H. javanica* var *podantha* (Molk.) Cl.; *H. nepalensis* Hk.; *H. polycephala* Wight. and Arn.

Family: Apiaceae

Distribution : Himalaya (Kashmir–Khasia), mountains of Malabar.

Part Used	:	Plant.
Reproductive Cycle	:	Rainy season.
Folk Use	:	Plant wrapped on hand and waist in frothy-stool and flatulence.
Active Constituents	:	Plant contains α-amyrin, lignoceric acid, and stigmasterol.
References	:	Asolkar *et al.* (1992), Barua *et al.* (2000), Bennet (1986).

Hymenachne amplexicaulis **(Rudge.) Nees.**

Syn: *H. acutigluma* (Steud.) Gilliland.; *H. myurus* Beauv.; *H. pseudointerrupta* C. Muell.; *Panicum acutioglumum* Steud.; *P. myurus* H.B and K.

Family: Poaceae

H. Name	:	*H.*: Dhamsiria.
Distribution	:	Marshy and wet places throughout the plains of Assam, Bengal, Bihar, Orissa.
Part Used	:	Pith.
Folk Use	:	Pith used for making garlands.
References	:	Ambasta (1986), Anonymous (1959), Bennet (1986), Ramachandran *et al.* (1986).

Ichnocarpus frutescens **(Linn.) Br.**

Syn: *Apocynum frutescens* Linn.; *Echites frutescens* (Linn.) Roxb.; *I. dasycalyx* Miq.; *I. radicans* Wall.

Family: Apocynaceae

Eng., H. and Sans. Names	:	*Eng.*: Black creeper. *H.*: Dudhilata, Kalidudhi, Siama-lata. *Sans.*: Paravalli, Sariva, Shariva, Syamalata.
Distribution	:	Throughout India upto 1,200m from W. Himalaya southwards to Tamil Nadu and Kerala.
Parts Used	:	Root. Stem.
Reproductive Cycle	:	September–April.
Folk Uses	:	A necklace of root said to induce sound sleep whereas that of stem pieces cure rheumatism. An amulet of stem pieces on the waist of children good against prolapsed anus.
Active Constituents	:	Triterpenoids, amyrin, friedelin, epifriedelin and β-sitosterol are the active constituents of stem.

References	:	Anonymous (1959), Girach and Aminuddin (1995), Khan and Khanum (2005), Rastogi and Mehrotra (1980-1984), Sen and Behera (2007), Singh (2000), Sood and Sudershna (2008).

Impatiens balsamina* Linn.

Syn: *I. lobbiana* Turez.; *I. salicifolia* Turez.

Family: Balsaminaceae

Eng., H. and Sans. Names	:	*Eng.*: Balsam weed, Garden balsam, Jewel weed. *H.*: Dupati, Gulmendhi. *Sans.*: Dushpatrijati, Tairini.
Distribution	:	Commonly cultivated in gardens throughout the tropical and subtropical parts of India.
Part Used	:	Leaves.
Reproductive Cycle	:	During rainy season.
Folk Uses	:	Used for staining palms; colouring skins and nails.
Active Constituents	:	2 hydroxy-1,4-naphthoquinone isolated from plant.
References	:	Ambasta (1986), Rastogi and Mehrotra (1960-1969), Siwakoti and Varma (1996).

***Impatiens gigantea* Edgew.**

Syn: *I. sulcata* Wall.

Family: Balsaminaceae

Eng. and H. Names	:	*Eng.*: Grooved balsam. *H.*: Khanchedargulmahdi.
Distribution	:	Temperate Himalaya.
Part Used	:	Whole Plant.
Reproductive Cycle	:	July–September.
Folk Use	:	Paste prepared by crushing the plant alongwith *Rumex scutatus* and some lichens in alcohol used as a substitute for nail polish.
References	:	Anonymous (1959), Bennet (1986), Sood *et al.* (2001).

***Indigofera astragalina* DC.**

Syn: *I. hirsuta* sensu Gamble.

Family: Fabaceae

Eng. Name	:	*Eng.*: Hairy indigo.
Distribution	:	Throughout plains, ascending to 1,200m in Kumaon Hills.

Part Used : Root.

Reproductive Cycle : September–November.

Folk Use : Root tied to the waist to cure fever.

References : Anonymous (1959), Bennet (1986), Rama Rao and Henry (1996), Sood and Sudershna (2008).

Indigofera cassioides Rottler *ex* DC.

Syn: *I. arborea* Roxb.; *I. elliptica* Roxb.; *I. glaucescens* Grah.; *I. jirahulia* Hamilt.; *I. pulchella* Roxb.; *I. purpurascens* Roxb.; *I. violacea* Roxb.

Family: Fabaceae

H. Names : *H.*: Hakna, Nil, Sakena.

Distribution : Moist evergreen forests throughout the Himalayan tract and hills upto 1,500m on Peninsular India.

Part Used : Fruit.

Reproductive Cycle : July–November.

Folk Use : Tribals give to their children for wearing a wreath made of one fruit in white thread as a charm against fever occurring at regular intervals.

References : Anonymous (1959), Bennet (1986), Singh and Narain (2008), Sood and Sudershna (2008), Sood *et al.* (2009).

Indigofera dosua Buch.–Ham. *ex* Don.

Syn: *D. heterantha* Wall. *ex* Brandis.; *I. gerardiana* Wall.; *I. polyphylla* DC.; *I. rubroviolacea* Dunn.; *I. stachyodes* Lindley.; *I. dosua* Ham. *ex* D. Don var *tomentosa* Baker.; *I. virgata* Roxb.

Family: Fabaceae

Eng. Name : *Eng.*: Himalayan Indigo.

Distribution : Found in H. P. and Temperate Himalaya:1,500-3,000m.

Part Used : Leaves.

Reproductive Cycle : May–July.

Folk Use : Used for tattooing.

References : Bennet (1986), Jain and Borthakur (1980), Sood *et al.* (2009).

Ipomoea aquatica Forsk.

Syn: *I. reptans* Poir.

Family: Convolvulaceae

Eng., H. and Sans. Names : *Eng.*: Swamp cabbage, Water spinach.

		H.: Kalmisag, Karmi, Patuasaga.
		Sans.: Kadambi, Kalaka, Kalambi, Kalambika, Nadika, Pechu.
Distribution	:	Abundant on moist soil or mud along the margins of stagnant streams, fresh water ponds and ditches in the plains of W. Bengal, Orissa and Assam.
Part Used	:	Flowers.
Reproductive Cycle	:	During rainy season.
Folk Use	:	Flowers used as adornments.
References	:	Anonymous (1959), Gogoi and Borthakur (1991), Khan and Khanum (2005), Kirtikar and Basu (1984).

Ipomoeapes-caprae **(Linn.) Sw.**

Syn: *Convolvulus pes-caprae* Linn.; *I. biloba* Forsk.; *I. maritime* R. Br.

Family: Convolvulaceae

Eng., H. and Sans. Names	:	*Eng.*: Goat's foot creeper.
		H.: Dopatilata.
		Sans.: Manmatha, Maravalli, Maryada, Sagara, Yugmapatra.
Distribution	:	Dry regions of Rajasthan and Andamans.
Parts Used	:	Flowers. Stem.
Reproductive Cycle	:	December–February.
Folk Uses	:	Onges women adorn their heads with flowers. Stem made into wristlet for marriage of Kumbhars of Poona.
Active Constituents	:	Flowers contain quercetin 3 glucoside, 3 galactoside and 3 (6″-O-acetyl) β glucopyranoside.
Reference	:	Awasthi and Goel (2000), Chandra (1997), Ferreira (1965), Kirtikar and Basu (1984), Prajapati *et al.* (2006), Rastogi and Mehrotra (1980-1984).

Ixora barbata **Roxb.**

Family: Rubiaceae

Part Used	:	Flowers.
Folk Use	:	Women wear flowers in hair.
Reference	:	Awasthi and Goel (2000).

Ixora brunnescens **Kurz.**

Family: Rubiaceae

Distribution	:	Andaman and Nicobar Islands.
Part Used	:	Flowers.
Folk Use	:	Women decorate their heads with flowers.
References	:	Awasthi and Goel (2000).

Jacaranda mimosifolia **D. Don**

Syn: *J. acutifolia* Humb. and Bonpl.; *J. ovalifolia* R. Br.

Family: Bignoniaceae

H. Name	:	*H.*: Jacaranda, Nili-gulmohar.
Distribution	:	Outer foothills of W. Himalaya.
Part Used	:	Flowers.
Folk Use	:	Flowers used for hair decoration by Kadukurubas.
Active Constituents	:	Constituents of flowers are 3 monoglycoside and 3,5-diglucoside of delphinidin.
References	:	Kshirsagar and Singh (2007), Rastogi and Mehrotra (1960-1969), Sood *et al.* (2009).

Jasminum angustifolium **(Linn.) Willd.**

Syn: *J. trifolium* Pers.; *J. vimineum* Willd.; *Nyctanthes angustifolia* Lamk.; *N. triflora* Burm., *N. viminea* Rets.

Family: Oleaceae

Eng., H. and Sans. Names	:	*Eng.*: Wild Jasmine. *H.*: Banmallika, Mwari. *Sans.*: Asphota, Kanamallika, Priya, Supuja, Vanamalli.
Distribution	:	Plant native to India.
Part Used	:	Flowers.
Folk Use	:	Flowers used for decorating hair and for making garlands by Soliga women.
Active Constituents	:	Fresh flowers yield indole.
Trade	:	Ban mallika.
References	:	Ambasta (1986), Khare (2004), Kirtikar and Basu (1984), Kshirsagar and Singh (2007), Sharma (2003).

Jasminum arborescens **Roxb.**

Syn: *J. latifolium* Roxb.; *J. roxburghianum* Wall.

Family: Oleaceae

Eng., H. and Sans. Names	:	*Eng.*: Tree jasmine. *H.*: Bela, Chameli, Mutabela, Navamallika, Saptala. *Sans.*: Nagamallika, Saptala, Varakunda.
Distribution	:	Upper Gangetic Plains, ascending to 1,000m on the Himalaya, Bengal, Central and S. India.
Part Used	:	Flowers.
Reproductive Cycle	:	April–July.
Folk Use	:	Flowers used for making garlands and adoring of hair by women.
Trade	:	Tree jasmine.
References	:	Chatterjee and Pakrashi (1997), Khan and Khanum (2005), Sen and Behera (2007), Sharma (2003).

Jasminum auriculatum **Vahl.**

Family: Oleaceae

Eng., H. and Sans. Names	:	*Eng.*: Needle flower jasmine. *H.*: Juhi, Jui. *Sans.*: Juthika, Mugdhee, Suchi-Mallika.
Part Used	:	Flowers.
Folk Use	:	Small fragnant flowers worn.
References	:	Ambasta (1986), Wealth (1959).

Jasminum flexile **Vahl.**

Syn: *J. azoricum* Linn.; *J. caudatum* Wall.

Family: Oleaceae

Part Used	:	Flowers.
Folk Use	:	Less fragnant flowers worn.
References	:	Bennet (1986), Wealth (1959), Watt (1889).

Jasminum grandiflorum **Linn.**

Syn: *J. aureum* Don.; *J. officinale* Linn.; *J. officinale* Linn. var. *grandifolorum* (Linn.) Stokes.; *J. officinale* fa *grandiflorum* (Linn.) Kobuski.

Family: Oleaceae

Eng., H. and Sans. Names	:	*Eng.*: Catalonian jasmine, Common jasmine, Common white garden, Jasmine, Spanish jasmine, White jasmine.

		H.: Chaluba, Chameli, Jati, Juhi.
		Sans.: Chameli, Malati, Mallika, Marati.
Distribution	:	Cultivated in gardens throughout India.
Part Used	:	Flowers.
Reproductive Cycle	:	April–September.
Folk Use	:	Women of S. India likes its flowers very much and used for making garlands and hair decoration. Women often ornament their coiffures and fore arms with garlands made from the white fragrant flowers of this shrub.
Active Constituents	:	Flowers rich in cis-3 hexenol, 2-vinylpyridine, indole, myrcene, linalool, a-terpineol, farnesol, eugenol, benzyl alcohol, methyl benzoate, benzyl cyanide, benzyl acetate, vanillin, methyl palmate and methyl linolate.
References	:	Bennet (1986), Dhiman (2003), Mehra *et al.* (1975), Rajendran and Aswal (2000), Raju (2000), Sood *et al.* (2009).

***Jasminum multiflorum* (Burm. f.) Andrews.**

Syn: *J. bracteatum* Wight.; *J. congestum* Wall.; *J. hirsutum* Willd.; *J. pubescens* (Retz.) Willd.; *Mogorium pubescens* Lamk.; *Nyctanthes multiflora* Burm.; *N. pubescens* Retz.

Family: Oleaceae

Eng., H. and Sans. Names	:	*Eng.*: Common jasmine, Downy jasmine, Musk jasmine.
		H.: Chameli, Kunda, Kundo, Kundphul.
		Sans.: Atimukha, Attahara, Kundah, Mahamoda, Makaranda, Manodna.
Distribution	:	Throughout India.
Part Used	:	Flowers.
Reproductive Cycle	:	November–April.
Folk Use	:	Used as adornments for making garlands and adorning of hair by women.
Active Constituents	:	Aerial parts yield seco-iridoid lactones-jasmolactones A, B, C, D and seco-iridoid glucoside–jasmultiside.
Trade	:	Safed chameli.
References	:	Bennet (1986), Gogoi and Borthakur (1991), Khan and Khanum (2005), Sharma (2003), Sood *et al.* (2009).

Jasminum sambac* (Linn.) Ait.

Syn: *J. fragrans* Salisb.; *J. pubescens* Wall.; *J. zambac* Roxb.; *Nyctanthes sambac* Linn.; *N. undulatum* Linn.

Family: Oleaceae

Eng., H. and Sans. Names	:	*Eng.*: Arabian jasmine, Lily jasmine, Sambac jasmine, Tuscan jasmine. *H.*: Banmallika, Champa, Mogra, Motia, Mugra. *Sans.*: Ananga, Atigandha, Bhupadi, Gauri, Mallika, Nava mullika.
Distribution	:	Cultivated throughout tropical and subtropical parts of the country.
Part Used	:	Flowers.
Reproductive Cycle	:	Hot and Rainy season.
Folk Use	:	Flowers used for making garlands and also worn on the ears and as hair ornament by women.
Active Constituents	:	Chief constituents of flowers are 9-deoxyjasminigenin, jasminin and 8, 9–dihydrojasmin.
References	:	Anonymous (1959), Dhiman (2003), Gogoi and Borthakur (1991), Khan and Khanum (2005), Mehra *et al.* (1975), Ramachandran *et al.* (1986), Rastogi and Mehrotra (1985-1989), Singh and Pandey (1998), Watt (1889).

***Jatropha gossypifolia* Linn.**

Family: Euphorbiaceae

Eng., H. and Sans. Names	:	*Eng.*: Bellyache bush, Tua-Tua. *H.*: Bherenda, Verenda. *Sans.*: Dravanti, Raktaeranda.
Distribution	:	Naturalized as a common weed in open degraded habitats throughout most of the plains of India.
Part Used	:	Root.
Reproductive Cycle	:	February–May.
Folk Use	:	A garland of pieces of its root around the neck of cattle believed to heal wounds.
Active Constituents	:	Roots contain macrocyclic diterpenoid jatrophone, 2α-hydroxyjatrophone, and 2 β hydroxyjatrophone.
References	:	Chatterjee and Pakrashi (1997), Jain (1991), Khan and Khanum (2005), Pal (1997), Rastogi and Mehrotra (1970-1979), Sood and Sudershna (2008), Sood *et al.* (2005).

Juglans regia Linn.

Syn: *J. arguta* Wall.; *J. kamaonia* C. DC.

Family: Juglandaceae

Eng., H. and Sans. Names	:	*Eng.*: Common walnut, European walnut, Himalayan walnut. *H.*: Akhor, Akhrot, Khor. *Sans.*: Akhoda, Akschoda, Akhrot.
Distribution	:	Cultivated in H.P., Kashmir, Kumaon and Khasia hills.
Part Used	:	Bark.
Reproductive Cycle	:	March–September.
Folk Use	:	Bark chewed as a substitute for lip-stick.
Active Constituents	:	Bark yields 5 hydroxy-1,4-napthoquinone, berberine, and 3,3′-bisjuglone, α-tetralone-(-).
Trade	:	Akahrot, Akhor, Akrut, Krot.
References	:	Dhiman (2003), Rastogi and Mehrotra (1985-1989), Sharma and Rana (2005), Sood *et al.* (2001, 2009).

Justicia adhatoda Linn.

Syn: *Adhatoda vasica* Nees.; *Adhathoda zeylanica* Medic.

Family: Acanthaceae

Eng., H. and Sans. Names	:	*Eng.*: Malabar nut, Vasaka. *H.*: Adalsa, Adarsa, Adusa, Arusha, Banasaa, Rusa, Vasaka. *Sans.*: Amalaka, Atarusha, Kanthiravi, Matrisinhi, Mrigendrani, Nasa, Panchamukhi, Ramrupaka, Sinhi, Sitakarni, Vasa, Vasaka, Vrisha.
Distribution	:	Throughout Indian plains.
Part Used	:	Wood.
Reproductive Cycle	:	December–May.
Folk Use	:	Small beads made from the wood of this evergreen shrub used for making rosaries.
Trade	:	Vasa, Vasaka.
References	:	Bennet (1986), Chauhan (1999), Jain (1991), Khan and Khanum (2005), Kirtikar and Basu (1984), Mehra *et al.* (1975), Sharma (2003), Sinha and Sinha (2001), Watt (1889).

Justicia betonica **Linn.**

Syn: ***Adhatoda betonica*** **Nees.;** ***J. betonica*** **var.** ***ramosissima*** **(Roxb.) Cl.;** ***J. orchroleuca*** **Blume.**

Family: Acanthaceae

Eng. and Sans. Names	:	*Eng.*: Rose spotted white flowered justice wort.
		Sans.: Sahacarah.
Distribution	:	Throughout India.
Part Used	:	Plant.
Reproductive Cycle	:	March–November.
Folk Use	:	Plant tied symbolically around the waist to keep away evil spirits.
References	:	Bennet (1986), Jain (1991), Sood *et al.* (2005), Varghese (1996).

Justicia gendarussa **Burm. f.**

Syn: ***Gendarussa vulgaris*** **Nees.**

Family: Acanthaceae

Eng., H. and Sans. Names	:	*Eng.*: Black purification plant.
		H.: Nili-nargandi, Udisanbhalu.
		Sans.: Bhuta-keshi, Gandharasa, Indrani, Kapika, Krishna-surasu, Marupatni, Nilamanjari, Nilasindhuka, Nilika, Pitasha, Sinduka, Vanaja.
Distribution	:	Wasteplaces, Hedges and Village shrubberies throughout India
Part Used	:	Leaves.
Reproductive Cycle	:	January–March.
Folk Use	:	Dried leaves kept inside pillow to relieve headache.
Active Constituents	:	Leaves exhibit β-sitosterol and alkaloids.
References	:	Arinathan and Mohan (2003), Khan and Khanum (2005), Kirtikar and Basu (1984), Rastogi and Mehrotra (1970-1979), Sood and Sudershna (2008).

Kaempferia angustifolia **Rosc.**

Family: Zingiberaceae

H. Names	:	*H.*: Kanjanbura, Mudunirbisha.
Distribution	:	Foot of E. Himalaya, Bengal.
Parts Used	:	Root. Seeds.

Folk Uses : Females of Karen tribe use the roots and seeds as a necklace for perfume and body ornamentation.

References : Ambasta (1986), Anonymous (1986), Jain (1991), Jain (1995), Mehra *et al.* (1975).

***Lablab purpureus* (Linn.) Sweet**

Syn: *Dolichos benghalensis* Jacq.; *D. lablab* Linn.; *D. lignosus* Linn.; *D. purpureus* Linn.

Family: Fabaceae

Eng., H. and Sans. Names : *Eng.*: Bonavist bean, Indian butter bean, Lablab bean.

H.: Bhatvas, Borboti, Lobia, Makhansim, Sem.

Sans.: Nispavah, Rajashimbi, Shimbi, Vallaka.

Distribution : Throughout India cultivated.

Parts Used : Branches. Shoots. Leaves.

Folk Uses : Sap obtained from the branches and shoots introduced below the skin with the help of thorns of *Acacia nilotica* in tattooing work by Bhils and Garasias. Leaves also used for tattooing.

References : Bennet (1986), Jain (1991), Prajapati *et al.* (2006), Singh and Pandey (1998), Sood *et al.* (2009).

***Lagenaria siceraria* Standl.**

Syn: *L. vulgaris* Ser.; *L. leucantha* Rusby.

Family: Cucurbitaceae

Eng., H. and Sans. Names : *Eng.*: Bottle Gourd, Calbash Gourd.

H.: Kaddu, Lauki, Tumri.

Sans.: Alabu.

Part Used : Fruit.

Folk Use : Gujarati Bhil girls wear neck of gourd fruit in ears.

References : Ambasta (1986), Francis (1984).

***Lannea coromandelica* (Houtt.) Merrill.**

Syn: *Dialium coromandelicum* Houtt.; *L. grandis* Dennst.; *Odina wodier* Roxb.

Family: Anacardiaceae

Eng., H. and Sans. Names : *Eng.*: Jhingam, Wodier.

H.: Jingan, Kaimil, Kimul, Mohin, Moyen.

Sans.: Ajasringi, Jhingini, Jingini, Manjari, Parvati, Pramodini.

Distribution : Throughout the hotter parts of India.

Part Used : Seeds.

Reproductive Cycle : February–June.

Folk Use : Ladies wear necklaces made of its seeds.

Trade : Wodier.

References : Anonymous (1962), Bennet (1986), Chatterjee and Pakrashi (1997), Khan and Khanum (2005), Prajapati *et al.* (2006), Singh and Pandey (1998).

Launaea procumbens **(Roxb.) Ramayya and Raja.**

Syn: *Chondrilla nudicaulis* Linn.; *L. nudicaulis* sensu Hk.f.; *Prenanthes procumbens* Roxb.

Family: Asteraceae

Distribution : Throughout India.

Parts Used : Leaves. Roots.

Reproductive Cycle : September–March.

Folk Use : Fresh leaves applied on fore head to reduce fever. Root pieces worn as garland to ward off evil spirits.

References : Bennet (1986), Chandra (1997), Jain (1991), Sinha and Sinha (2001), Sood *et al.* (2005).

Lawsonia alba **Lamk.**

Syn: *L. inermis* (Linn.) Linn.; *L. purpurea* Lamk.; *L. spinosa* Linn.

Family: Lythraceae

Eng., H. and Sans. Names : *Eng.*: Camphire, Cypress shrub, Egyptian privet, Henna, Jamaica, Tree mignonette.

H.: Hena, Mehndi.

Sans.: Kuravaka, Mehndi, Mendiki, Ranjaka, Timira, Yavaneshta.

Distribution : Throughout India; cultivated in drier parts of Peninsula, Panjab, Gujarat, M.P. and Rajasthan.

Parts Used : Leaves. Flowers.

Reproductive Cycle : April–July.

Folk Uses : A crimson red dye extracted from leaves used for colouring palm, sole and nails of bride before marriage. Garlands made from the flowers used by N. Indian women to adorn their hair.

Active Constituents : Gallic acid, 2 hydroxy-1,4-napthoquinone,1,4-dihydroxynapthalene, 1,4-napthoquinone, fraxetin,

		scopotelin and esculetin are the active constituents of leaves.
Trade	:	Henna.
References	:	Anonymous (1962), Bennet (1986), Jain (1991), Mehra *et al.* (1975), Rastogi and Mehrotra (1980-1984), Sharma (2003), Sood *et al.* (2005), Vihari (1995).

Leea macrophylla **Roxb.** ***ex*** **Hornem.**

Syn: *L. angustifolia* Laws.; *L. cinera* Laws.; *L. diffusa* Laws.; *L. integrifolia* Roxb.

Family: Leeaceae

H. and Sans. Names	:	*H.*: Dholsamudra, Samudraka.
		Sans.: Dholasamudrika, Samudraka.
Distribution	:	Throughout hotter parts of India.
Part Used	:	Seeds.
Reproductive Cycle	:	January–June.
Folk Use	:	Seeds wrapped in a cloth and tied around the neck of child to cure stomach pain.
References	:	Kirtikar and Basu (1984), Pradhan and Badola (2008).

Leucaena leucocephala **(Lamk.) Wit.**

Family: Fabaceae

Eng. Name	:	*Eng.*: White popinac.
Parts Used	:	Fruits. Seeds.
Folk Use	:	Fruits and seeds used for ornament purpose.
Reference	:	Mehra *et al.* (1975).

Leucas aspera **Spreng.**

Syn: *L. dimidiata* Benth.; *L. plukeneti* Roth.; *Phlomis aspera* Willd.; *P. esculenta* Roxb.; *P. obliqua* Ham.; *P. plukenetti* Roth.

Family: Lamiaceae

Eng., H. and Sans. Names	:	*Eng.*: White dead nettle.
		H.: Chot-hal-kusa, Chhota-italkusa.
		Sans.: Dronapushpi.
Distribution	:	Plains of India.
Parts Used	:	Flowers. Root.
Reproductive Cycle	:	July–January.

Folk Uses	:	Flowers used as adornments. Roots collected in the early morning on Saturday or Sunday worn to cure cough, asthma and stomach disorders.
Active Constituents	:	A new triterpenoid lactone–leucolactone isolated from roots.
References	:	Anonymous (1962), Bennet (1986), Gogoi and Borthakur (1991), Jain *et al.* (1991), Sen and Behera (2007), Sood and Sudershna (2008), Sood *et al.* (2005).

Leucas cephalotes **(Roth.) Spreng.**

Syn: *L. capitata* Desf.; *Phlomis cephalotes* Roth.

Family: Lamiaceae

H. and Sans. Names	:	*H.*: Deldona, Dhurpisag, Goma, Guma, Motapati. *Sans.*: Chhatraka, Chhatrani, Drona pushpin, Palindi.
Distribution	:	Panjab, Bengal, Assam, Himalaya, Tamil Nadu.
Part Used	:	Stem.
Reproductive Cycle	:	Almost throughout year.
Folk Use	:	Tribals give a wreath made of a piece of stem for wearing to children to cure skin diseases.
Trade	:	Dronpushpi.
References	:	Anonymous (1962), Chauhan (1999), Hembrom (1996), Jain *et al.* (1991), Khan and Khanum (2005), Singh and Narain (2008).

Licuala peltata **Roxb.**

Family: Arecaceae

Distribution	:	N. Bengal, lower Sikkim, Assam, Khasia, Naga Hills, Orissa and Andaman Islands.
Part Used	:	Leaves.
Folk Use	:	Head bands, necklaces, armlets and waist girdles of its leaves worn by the Jarawa tribe women of Andaman.
References	:	Anonymous (1962), Chandra (1997).

Linum usitatissimum **Linn.**

Syn: *L. crepitans* (Boenn.) Dumort.; *L. humile* Mill.; *L. trinervium* Roth.

Family: Linaceae

Eng., H. and Sans. Names	:	*Eng.*: Flax, Linseed. *H.*: Alsi, Tisi. *Sans.*: Atasi.

Distribution : Cultivated throughout subtropical and temperate regions of India upto 2,000 m.

Part Used : Stem.

Reproductive Cycle : December–March.

Folk Use : Pieces of stem strung and used as necklace and amulets in N. India.

References : Ambasta (1986), Mehra *et al.* (1975), Sood *et al.* (2009), Watt (1889).

Litsea monopetala **(Roxb.) Pers.***

Syn: *L. polyantha* Juss.

Family: Lauraceae

H. Names : *H.*: Kakuri, Katmarra, Meda, Patoia.

Distribution : Western outer Himalaya.

Part Used : Bark.

Folk Use : Bark tied on the fractured bones to get relief.

References : Ambasta (1984).

Livistonia jenkinsiana **Griff.**

Family: Arecaceae

Eng. Name : *Eng.*: Assam palm.

Distribution : Throughout Assam.

Parts Used : Leaves. Shoots.

Reproductive Cycle : May–December

Folk Use : Used as adornments.

References : Gogoi and Borthakur (1991), Watt (1972).

Livistonia speciosa **Kurz.**

Family: Arecaceae

Distribution : Tropical forests of Chittagong.

Part Used : Leaves.

Folk Use : Leaves used as adornments.

References : Gogoi and Borthakur (1991), Watt (1972).

Lobelia nicotianifolia **Roth.** ***ex*** **Roem. and Sch.**

Syn: *L. beddomeana* Wimmer; *L. colorata* Wall.; *L. erecta* Hk.f. and Thoms.; *L. pyramidalis* Wall.; *L. rosea* Wall.; *Rapuntium nicotianaeplium* Presl.

Family: Campanulaceae

Eng., H. and Sans. Names	:	*Eng.*: Wild tobacco. *H.*: Nala, Narasala. *Sans.*: Bibhishana, Chhidrant, Devanala, Dhamana, Kichaka, Mala.
Distribution	:	Hilly regions of Maharashtra, Tamil Nadu. and M.P.
Part Used	:	Stem bark.
Folk Use	:	A piece of stem bark placed on the cavity in tooth to check pain.
Active Constituents	:	Stem bark contains lobeline.
Trade	:	Wild tobacco.
References	:	Bennet (1986), Jain *et al.* (1991), Kirtikar and Basu (1984), Rastogi and Mehrotra (1970-1979).

Luffa acutangula **(Linn.) Roxb.**

Syn: *Bryonia cheirophylla* Wall.; *Cucumis acutangulus* Linn.; *L. aegyptica* Mill.; *L. amara* Roxb.; *L. clavata* Roxb.; *L. kleinii* W. and A.; *L. pentandra* Roxb.; *L. petola* Seringe.; *L. racemosa* Roxb.; *L. satpatia* Wall.; *L. umbellata* (Klein. *ex* Willd.) Roem.; *Momordica cylindrica* Linn.; *M. luffa* Linn.

Family: Cucurbitaceae

Eng., H. and Sans. Names	:	*Eng.*: Gourd towel, Sponge gourd, Vegetable sponge, Washrag sponge. *H.*: Ghiaturai, Jinga, Nenua, Purula. *Sans.*: Aibhi, Brihatkoshtaki, Ghoshaka, Hastighosha, Hastiparna, Mahaphala, Sapitaka.
Distribution	:	Throughout India.
Parts Used	:	Roots. Leaf.
Reproductive Cycle	:	June–December.
Folk Uses	:	Roots alongwith that of *Mucuna pruriens* tied on waist of males on Saturday to check premature ejaculation. Whole leaf tied on the breast to avoid over lactation after childbirth.
Active Constituents	:	Roots yield cucurbitacins B and D.

References : Anonymous (1962), Bennet (1986), Sarangi and Sahu (2004), Sen *et al.* (2008), Sood and Sudershna (2008), Sood *et al.* (2009).

Lygodium flexuosum (Linn.) Sw.

Syn: *L. pinnatifidum* Sw.

Family: Schizaeaceae

Eng. and H. Names : *Eng.*: Sinuate lygodium.

H.: Bhutraj.

Distribution : Throughout India.

Part Used : Roots.

Reproductive Cycle : February–December.

Folk Uses : Roots hung around the neck of cattle to repel the parasites. It is believed that the touch of the roots cures epilepsy and hysteria.

Active Constituents : Quercetin and 3,5,7,3′,4′-pentahydro-xyflavone-3-o-rutinoside have been isolated from roots.

References : Jain *et al.* (1991), Molla and Roy (1996), Rastogi and Mehrotra (1985-1989), Sahoo and Bahali (2003), Sood and Sudershna (2008), Sood *et al.* (2005), Varghese (1996).

Lygodium japonicum (Thunb.) Sw.

Family: Schizaeaceae

Distribution : Kashmir eastwards to Lushai hills rarely in W. Ghats upto 2,000m.

Part Used : Whole Plant.

Folk Use : An amulet of this fern tied around the neck of patient suffering from a kind of fever known among the Mishings as 'Bhalukjor' probably malaria.

Active Constituents : An antheridiogen isolated and characterized from the plant as methyl ester of new gibberellin GA 73.

References : Anonymous (1962), Das (1997), Rastogi and Mehrotra (1985-1989).

Madhuca indica J.F. Gmelin

Syn: *Bassia latifolia* Roxb.; *M. latifolia* (Roxb.) Mac. Bridge.; *M. longifolia* Koen.

Family: Sapotaceae

Eng., H. and Sans. Names : *Eng.*: Indian butter tree, Mowra butter tree, South Indian mahua.

		H.: Mahuwa, Mauro, Mawda, Mouo, Mowa, Muall, Muwa.
		Sans.: Madhuka.
Distribution	:	Throughout W. Bengal, Bihar, Orissa, M.P., Panjab and U.P.
Parts Used	:	Flowers. Wood.
Reproductive Cycle	:	March–May.
Folk Uses	:	Ho girls wear flower garlands in their hair as adornment. Bride wears bead cut from its wood.
Active Constituents	:	Flowers contain–Polysaccharide, PS-AI, PS-AII, galactose, glucose, arabinose and glucuronic acid are the active constituents of flowers.
Trade	:	Madhua.
References	:	Anonymous (1962), Chatterjee and Pakrashi (1997), Jain (1991), Khan and Khanum (2005), Majumdar (1950), Rao *et al.* (1996), Rastogi and Mehrotra (1980-1984), Sharma (2003), Singh and Pandey (1998), Sood *et al.* (2005).

Magnolia campbellii **Hk. f. and Th.**

Family: Magnoliaceae

Eng. Name	:	*Eng.*: Red magnolia.
Distribution	:	E. Himalaya, Manipur.
Part Used	:	Flowers.
Folk Use	:	Used for hair decoration
References	:	Anonymous (1962), Saklani and Jain (1994).

Mallotus peltatus **(Geis.) Muell. Arg.**

Syn: *Aleurites peltatus* Geis.; *M. acuminatus* (Bl.) Muell.

Family: Euphorbiaceae

Part Used	:	Leaves.
Folk Use	:	Onge women tie fresh green leaves at their genital parts to reduce the menstrual pains.
References	:	Awasthi and Goel (2000), Bennet (1980).

Mallotus philippensis **(Lam.) Muell.-Arg.***

Syn: *Croton cascarilloides* Rauesch; *C. coccineus* Vahl.; *C. distanus* Wall.; *C. montanus* Willd.; *C. philipensis* Lamk.; *C. punctatus* Retz.; *Rottlera affinis* Hassk.; *R. aurantiaca* Hook. and Arn.; *R. montana* Wall.; *R. tinctoria* Roxb.

Family: Euphorbiaceae

Eng., H. and Sans. Names : *Eng.*: Kamala tree, Monkey face tree.

H.: Kamala, Kambila, Kambhal, Kamud, Rohini, Ruin, Rulu, Rutin.

Sans.: Kapila, Kampilla, Rechanaka.

Distribution : Throughout tropical regions of India, in the Himalaya upto 1,500m and southwards upto Kerala.

Part Used : Glandular Hair of Fruit.

Reproductive Cycle : November–April.

Folk Use : Glandular hair from fruits yield a dye called 'kamala' powder used as 'sindhur' or 'kumkum' (vermilion) by Indian women.

Active Constituents : Rottlerin is the main colouring pigment.

References : Ambasta (1986), Chauhan (1999), Sood and Thakur (2004), Sood *et al.* (2009).

Mangifera indica **Linn.**

Syn: *M. domestica*Gartn.

Family: Anacardiaceae

Eng., H. and Sans. Names : *Eng.*: Cuckoo's joy, Mango Tree, Spring Tree.

H.: Am, Amb.

Sans.: Alipriya, Amra, Chotu, Manjari.

Distribution : Throughout N. E. region, tropical Himalaya (Kumaon–Bhutan hills), valleys of Bihar, Khasia Hills, Mayanmar and W. Peninsula.

Part Used : Leaf.

Reproductive Cycle : February–July.

Folk Use : Leaf forms an important part of the *kangan* (wrist band) ornament tied to wrists of the wedding couple.

Active Constituents : Gallic acid, ethyl ester, chinomin, quercitin and hyperin (leaves).

References : Ambasta (1986), Birwood (1880), Rastogi and Mehrotra (1998, 2001), Reddy (1956).

Manilkara littoralis **(Kurz.) Dub.**

Syn: *Mimusops indica* Kurz. not of A.DC.; *M. littoralis* Kurz.

Family: Sapotaceae

Eng. Name	:	*Eng.*: Andaman bullet wood.
Distribution	:	Andaman and Nicobar Islands.
Parts Used	:	Flowers. Bark. Leaves
Folk Uses	:	Bark strip tied to the Onge women after delivery to restore the normality of the abdomen. Flowers used to prepare hair bands, necklaces and armlets. Leaves worn on Andamese women's belts
Active Constituents	:	Gallic acid, ellagic acid, methyl gallate, quercetin and polymeric tannin characterized from bark.
References	:	Anonymous (1962), Chakrabarty and Balarkishnan (2003), Dagar and Dagar (2000), Radcliffe-Brown (1922), Rastogi and Mehrotra (1970-1994).

Marsilea minuta **Linn.**

Family: Marsileaceae

H. and Sans. Names	:	*H.*: Chaupatira. *Sans.*: Sunisannak.
Distribution	:	Throughout plains of India.
Part Used	:	Whole Plant.
Folk Use	:	Plant tied on the waist of children during stomach pain.
Active Constituents	:	Plant contains triterpene marsileagenin A, β-sitosterol, methylamine, hentriacontan-6-ol, C-glucosyl flavones and C–glucosylxanthne.
References	:	Chatterjee and Pakrashi (1997), Jain and Patole (2001), Rastogi and Mehrotra (1970-1979).

Melastoma malabathricum **Linn.**

Syn: *M. normale* D.Don.; *M. polyanthum* Blume.

Family: Melastomataceae

Eng., H. and Sans. Names	:	*Eng.*: Indian rhododendron. *H.*: Phutk. *Sans.*: Tinisah.
Distribution	:	Throughout India, very abundant from sea level upto 2,000m

Part Used	:	Flowers.
Reproductive Cycle	:	October–March.
Folk Use	:	Flowers used for decorating hair and giving deep blue colour to lips and tongue.
Trade	:	Malabar gooseberry.
References	:	Anonymous (1962), Bhargava (1997), Gogoi and Borthakur (1991), Hooker (1897), Khan and Khanum (2005), Saklani and Jain (1994).

Melia azedarach Linn.

Syn: *M. bukayan* Royle.; *M. sempervirens* Sw.

Family: Meliaceae

Eng., H. and Sans. Names	:	*Eng.*: Barbados lilac, Bead tree, Indian lilac, Persian lilac. *H.*: Bakain, Bakarja, Deikna, Drewk, Mahanimbo. *Sans.*: Brihan nimba, Mahanimba, Mahadroksha, Vishamushtika.
Distribution	:	Wild in the forests of hotter parts of India.
Part Used	:	Fruit.
Reproductive Cycle	:	May–October.
Folk Use	:	Stones of the fruits strung into necklaces and rosaries and worn as a charm against diseases.
Active Constituents	:	Constituents isolated from fruits are 6 tetranortri-terpenoids ochinolide A, ochinolide B, nimbolidin A, nimbolidin B, meliatoxins A1,A2, B1, and B2.
Trade	:	Bead tree.
References	:	Anonymous (1962), Dhiman (2003), Rastogi and Mehrotra (1980-1984), Sharma (2003), Sood *et al.* (2009).

Melothria heterophylla (Lour.) Cogn.

Syn: *Solena amplexicaulis* (Lamk.). Gandhi.; *S. heterophylla* (Lour.) Cogn.; *Zehneria umbellata* Thw.

Family: Cucurbitaceae

Eng., H. and Sans. Names	:	*Eng.*: Diversifolius melothira. *H.*: Amantmul, Tarali. *Sans.*: Sukakandah.
Distribution	:	Throughout India.
Part Used	:	Root.

Reproductive Cycle	:	May–October.
Folk Use	:	Pieces of fresh root alongwith that of *Plumbago indica* and rhizome of *Curcuma domestica* worn around neck for 15-20 days to reduce jaundice.
Active Constituents	:	Lignoceric, tricosanoic and behenic acids are the active constituents of roots.
References	:	Anonymous (1962), Pandey *et al.* (1996), Rastogi and Mehrotra (1970-1979), Sood *et al.* (2009).

Melothria maderaspatana (Linn.) Cogn.

Syn: *Bryonia scabrella* L.f., *Cucumis maderaspatana* Linn., *Mukia maderaspatana* (Linn.) Roem., *M. scabrella* Arn.

Family: Cucurbitaceae

H. Names	:	*H.*: Agumaki, Bilari.
Distribution	:	Throughout India.
Part Used	:	Fruits.
Reproductive Cycle	:	July–December.
Folk Use	:	Fruits bandaged over swollen finger in the treatment of paronichia.
References	:	Anonymous (1962), Katewa and Sharma (1998), Kirtikar and Basu (1984).

Mesua ferrea Linn.

Syn: *M. pedunculata* Wight.; *M. nagana* Gard.; *M. salicina* Planch. and Trian.; *M. sclerophylla* Thwaites.; *M. speciosa* Chois.; *M. thwaitesii* Planch. and Triana.

Family: Clusiaceae

Eng., H. and Sans. Names	:	*Eng.*: Ironwood of Assam, Srilanka ironwoood. *H.*: Nagesr, Nagkesar. *Sans.*: Bhujangakhya, Hema, Kanchana, Kesara, Nagkeshara, Nagapuspah, Pinjara.
Distribution	:	Mostly in Eastern and S. India.
Parts Used	:	Flowers. Leaves.
Reproductive Cycle	:	February–September.
Folk Uses	:	Bunches of the delicately coloured young leaves used for decorating the hair styles of women in Assam. The Assamese people also wear the flowers in their ear perforations.

Active Constituents	:	Leaves contain a new flavanone glycoside mesuein.
Trade	:	Nagkesur.
References	:	Bennet (1986), Dhiman (2003), Gogoi and Borthakur (1991), Jain (1991), Mehra *et al.* (1975), Prajapati *et al.* (2006), Rastogi and Mehrotra (1985-1989), Saklani and Jain (1994), Sinha and Sinha (2001), Watt (1972).

Michelia champaca Linn.

Syn: *M. aurantiaca* Wall.; *M. doldsopa* Ham.; *M. excela* Bl.; *M. rheedi Wight.*; *M. rufinervis* DC.

Family: Magnoliaceae

Eng., H. and Sans. Names	:	*Eng.*: Champaca, Golden champa, Yellow champa. *H.*: Champ, Champa, Champac, Champaca. *Sans.*: Anjana, Bhringmohi, Champakah, Katu, Kumara.
Distribution	:	E. Himalaya. N. E. India. W. Ghats.
Part Used	:	Flowers.Wood.
Reproductive Cycle	:	April–October.
Folk Uses	:	Flowers worn by the ladies of S. India for their hairdo. The sweet–scented yellow flowers of this evergreen tree also worn in turbans and coiffures. Wood cut into beads for adornment.
Active Constituents	:	Mono and sesquiterpenes isolated from flowers.
Trade	:	Champak.
References	:	Anonymous (1962), Bennet (1986), Chatterjee and Pakrashi (1997), Dastur (1964), Dhiman (2003), Mehra *et al.* (1975), Raju (2000), Rastogi and Mehrotra (1970-1979), Saklani and Jain (1994), Sood *et al.* (2009).

Millettia fruticosa Benth. *ex* Baker.

Syn: *Amerimim fasciculatum Hamilt.*; Otosema fruitcosa Benth.; *Pongamia fruticosa* Grah.; *Robinia fruticosa* Roxb.

Family: Fabaceae

Distribution	:	Forests of Bengal, Assam.
Part Used	:	Whole Plant.
Folk Use	:	Touching the plant brings good luck during tug of war (Thingreira-khangakhun).
References	:	Hooker (1897), Saklani and Jain (1994), Sood *et al.* (2005).

***Mimosa pudica* Linn.**

Syn: *M. hispidula* Kunth.

Family: Fabaceae

Eng., H. and Sans. Names	:	*Eng.*: Humble plant, Sensitive plant. *H.*: Chui-Mui, Lajalu, Lajwanti, Sharmpate. *Sans.*: Anjalikaraka, Lajjahah, Lajjalu, Mahabhita, Mamaskari, Tamra, Vashini.
Distribution	:	Hotter parts throughout India.
Parts Used	:	Leaves. Root. Stem. Whole Plant.
Reproductive Cycle	:	Throughout year.
Folk Uses	:	Roots collected in the early morning on Saturday/ Sunday and tied around the neck of children to cure cough and for warding off the snakes. 3 cm long piece of its stem tied as an amulet on Sunday morning after reciting oracles in the neck of child suffering from fever. Plant also kept below the pillow while sleeping to get sound sleep and curing insomania. Leaves tied in a knot of worn cloth effective in inducing peaceful sleep by Bettakurubas.
Active Constituents	:	Properties of root attributed to β-sitosterol, D-pinitol, norepinephrine, mimosine, crocetin dimethyl ester and tannin.
Trade	:	Sensitive plant.
References	:	Aminuddin and Girach (1991), Chatterjee and Pakrashi (1997), Cyril Nayagam and Pushparaj (2000), Khan and Khanum (2005), Kshirsagar and Singh (2001), Sen and Behera (2007), Sinha and Sinha (2001).

***Mimusops elengi* Linn.**

Family: Sapotaceae

Eng., H. and Sans. Names	:	*Eng.*: Bullet wood, Spanish cherry, West India Medlar. *H.*: Bakul-mulsari, Maulsarau, Maulser, Mulsari. *Sans.*: Anjalikaraka, Lajjahah, Lajjalu, Mahabhita, Mamaskari, Tamra, Vashini.
Distribution	:	Throughout tropical India.
Parts Used	:	Flowers. Seeds.
Reproductive Cycle	:	March–June.
Folk Use	:	Flowers and seeds used for making garlands and hairdo.

Active Constituents	:	These are D-mannitol, β-sitosterol, and its glucosides (flowers).
Trade	:	Indian Medlar Tree.
References	:	Anonymous (1962), Jain (1991), Khare (2004), Rastogi and Mehrotra (1970-1979), Saklani and Jain (1994), Sharma (2003), Watt (1889).

Mirabilis jalapa* Linn.

Syn: *M. multiflora* (Torr.) Gray.

Family: Nyctaginaceae

Eng., H. and Sans. Names	:	*Eng.*: Four O'clock flower, Marvel of Peru. *H.*: Gulabash, Gulabbas, Guleaabbas, Kishna-keli. *Sans.*: Krishna-keli, Sandhyakali, Tridosha, Trisondhi.
Distribution	:	Naturalised widely in India.
Parts Used	:	Roots. Flowers. Seeds.
Reproductive Cycle	:	During summer and Rainy Season.
Folk Uses	:	Seeds and flowers worn as adornments. Pieces of its root tied to waist for smooth delivery and relief from body pains and to the neck for expelling the placenta from the uterus of domestic animals.
Active Constituents	:	Roots exhibited linear D glucan, stigmasterol and β-sitosterol.
References	:	Chatterjee and Pakrashi (1997), Jain *et al.* (1991), Mehra *et al.* (1975), Molla and Roy (1996), Rastogi and Mehrotra (1985-1989), Saklani and Jain (1994), Sinha and Sinha (2001), Sood and Sudershna (2008), Sood *et al.* (2005).

***Molineria prainiana* Deb.**

Syn: *Curculigo prainiana* (Deb.) Bennet and Raizada.

Family: Liliaceae

Part Used	:	Leaves.
Folk Use	:	Fresh leaves tied over the lower abdomen of pregnant lady for relieving of labour pains and quick delivery.
References	:	Rawat and Chowdhury (1998), Saklani and Jain (1994).

***Momordica charantia* Linn.**

Syn: *Cucumis africanus* Bot.; *M. humilis* Wall.; *M. muricata* DC.; *M. senegalensis* Lamk.

Family: Cucurbitaceae

Eng., H. and Sans. Names	:	*Eng.*: Balsam pear, Bitter gourd, Carilla fruit.

H.: Karela, Kareli, Poti-kakar, Uchche.

Sans.: Brihadavalli, Karavellah, Sushavi.

Distribution : Cultivated Throughout India.

Part Used : Leaf.

Reproductive Cycle : July–October.

Folk Use : Leaf tied on the boils for suppuration by Soligas.

Active Constituents : Leaves reported to contain mordicines I, II and III, cucurbitane, octacosane, 1 triacontanol, 7 stigmasten 3 β-ol and α phytosphingosine.

Trade : Bitter Gourd.

References : Khan and Khanum (2005), Kshirsagar and Singh (2007), Rastogi and Mehrotra (1980-1984), Sood and Sudershna (2008), Sood *et al.* (2009).

Mucuna pruriens **(Linn.) DC.**

Syn: *Carpopogon niveum* Roxb.; *Caryopogon pruriens* Roxb.; *Dolichos pruriens* Linn.; *M. nivea* (Roxb.) DC.; *M. prurita* Hk.; *M. utilis* Wall.; *Stizolobium pruriens* Linn. Medic.

Family: Fabaceae

Eng., H. and Sans. Names : *Eng.*: Cowhage, Cowitch.

H.: Erka, Goncha, Kaunch, Kivachh.

Sans.: Adhyanda, Ajavha, Arshabhi, Durabhigraha, Kapiprabha.

Distribution : Throughout India.

Part Used : Root.

Reproductive Cycle : July–November.

Folk Uses : Roots collected in early morning on Saturday or Sunday and tied to the arm of male partner for prolonged sexual intercourse. A piece of root also tied over wrist in dropsy.

Active Constituents : Roots rich in prurieninine and prurienidine.

Trade : Kaunch beej.

References : Bennet (1986), Chauhan (1999), Dhiman (2007), Sen and Behera (2007), Sinha and Sinha (2001), Sood *et al.* (2009).

Murraya koenigii Spreng.

Syn: *Bargera koenigii* Linn.

Family: Rutaceae

Eng., H. and Sans. Names :	*Eng.*: Curry Leaf Tree. *H.*: Kathnim, Kurry Patta, Mitha Neem. *Sans.*: Krishna-nimba, Nimba-pattram, Surabhininiba.
Distribution :	Throughout India, ascending to 1,650m.
Part Used :	Seeds.
Reproductive Cycle :	March–June.
Folk Use :	The Chenchus string seeds on necklaces.
References :	Ambasta (1986), Furer-Haimendorf (1943), Sood *et al.* (2005, 2009).

Murraya paniculata (Linn.) Jack.

Syn: *Chalcas paniculata* Linn.; *C. sumatrana* Roem.; *M. exotica* Linn.; *M. sumatrana* Roxb.

Family: Rutaceae

Eng., H. and Sans. Names :	*Eng.*: Honey bush, Orange jassamine. *H.*: Bibsar, Marchula, Juti, Kamini. *Sans.*: Ekangi, Muramamsi.
Distribution :	Outer Himalaya from the Jamuna eastwards, Assam, Satpura range.
Part Used :	Flowers.
Reproductive Cycle :	April–January.
Folk Use :	Flowers used for hair decoration by Manipur people.
Active Constituents :	Flowers contain scopolin, scopoletin and glucose.
Trade :	Marchula.
References :	Anonymous (1962), Bennet (1986), Kirtikar and Basu (1984), Rastogi and Mehrotra (1970-1979), Saklani and Jain (1994), Sharma (2003).

Musa glauca Roxb.

Syn: *Ensete glaucum*.

Family: Musaceae

Distribution :	Cultivated in N.E. India.
Part Used :	Seeds.

Folk Use : Seeds made into beads and used as necklace by children against convulsions associated with fever.

Reference : Lalramnghinglova (2003).

Musa paradisiaca **Linn.***

Syn: *M. sapientum* Linn.

Family: Musaceae

Eng., H. and Sans. Names : *Eng.*: Adam's Fig, Banana, Fig of India, Plantain.

H.: Amrit, Banana, Kachkula, Kela, Maozkula.

Sans.: Alabu, Ambusara, Kadli, Mocha, Nisara, Rambha, Sakritphala.

Distribution : Cultivated throughout India mainly in Assam, A.P., Bihar, Kerala, M.P., U.P. and Tamil Nadu.

Parts Used : Leaves. Pseudostem. Root. Seeds.

Reproductive Cycle : May–July.

Folk Uses : Seeds worn as beads by the Kacha Nagas.The witch doctor usually wears a garland of its leaves while removing bad spirits from a man. Wearing a piece of the membranous outer covering of the pseudostem cures skin allergies due to catterpillars. A garland of root piece hung around the neck during early morning on a Sunday for curing goitre. Also, roots tied on the arm in yellow cloth to get relief from the ill effect of Jupiter planet.

Trade : Jangli Kele.

References : Dagar and Dagar (1996), Dhiman (2003), Hembrom (1996), Hutton (1921a), Sharma (2003), Sharma *et al.* (2003), Sinha and Sinha (2001), Sood *et al.* (2005).

Mussaenda frondosa **Linn.**

Syn: *M. corymbosa* Roxb.; *M. flavescens* Ham.; *M. formosa* Linn.; *M. frendosa* var. *glabrata Hk .f.*; *M. glabrata* (Hk.f.) Hutch. *ex* Gamble.; *M. villosa* Wall.; *M. zeylanica* Burm.

Family: Rubiaceae

Eng., H. and Sans. Names : *Eng.*: White lady.

H.: Asari, Bebina, Bedina.

Sans.: Nagavalli, Shrivati.

Distribution : N. E. Himalaya. Bengal. S. India.

Part Used : Leaves.

Reproductive Cycle : During rainy season.

Folk Use : Leaves considered as a charm against demons.

References : Anonymous (1962), Dhiman (2003), Sood and Sudershna (2008).

Myristica fragrans **Houtt.**

Family: Myristicaceae

Eng., H. and Sans. Names : *Eng.*: Nutmeg Tree.

H.: Jaiphal, Japatri, Jotri, Jayapatri.

Sans.: Jatiphala.

Part Used : Arils.

Folk Use : Arils strung on garlands in Thanjaver, Tamil Nadu.

References : Ambasta (1986), Francis (1984).

Nannorhops ritchieana **H. Wendl.**

Family: Arecaceae

Eng. Name : *Eng.*: Mazari Palm.

Distribution : N. India.

Part Used : Seeds.

Folk Use : Seeds strung for rosaries.

References : Ambasta (1986), Dastur (1964).

Naravelia zeylanica **(Linn.) DC.**

Syn: *Atragene zeylanica* Linn.

Family: Ranunculaceae

Sans. Name : *Sans.*: Dhanavalli.

Distribution : Tropical forests of E. Himalaya, Assam, W. Bengal, Bihar, greater parts of Deccan Peninsula.

Parts Used : Stems. Leaves.

Reproductive Cycle : Flowering (rainy season) and Fruiting (winters).

Folk Use : Stem alongwith its leaves tied around the forehead to cure headache by Peniyeravas, Malekudiyas and Girijanas.

References : Chatterjee and Pakrashi (1997), Kshirsagar and Singh (2007).

Naringi crenulata **(Roxb.) Nicolson.**

Syn: *Hesperethusa ambigua* Roem.; *H. crenulata* (Roxb.) M. Roem.; *Limonia acidissima* auct. (non Linn.); *L. crenulata* Roxb.

Family: Rutaceae

Sans. Name	:	*Sans.*: Bilvaparni.
Distribution	:	Dry hills of N.W. Himalaya, Shimla, Kumaon; ascending to 1,700m.
Part Used	:	Leaves.
Folk Use	:	Amulet of its leaves tied to arms to ward off evil spirits.
Active Constituents	:	Leaves contain luvangetin, limonin, marmesin, xanthotoxin, sitosterol and its glucoside.
References	:	Bennet (1986), Jain (1991), Rastogi and Mehrotra (1960-1989), Sood and Sudershna (2008), Sood *et al.* (2005).

Nelumbo nucifera **Gaertn.**

Syn: *Cyamus mysticus* Salisb.; *C. nelumbo* Smith.; *N. indica* Poir.; *N. speciosum* Willd.; *Nymphaea asiaticum* Rich.; *N. nelumbo* Linn.

Family: Nymphaeaceae

Eng., H. and Sans. Names	:	*Eng.*: Chinese water lily, Egyptian bean, The Indian Lotus, The Sacred Lotus, Pythagorean Bean. *H.*: Ambuj, Kamal, Kanval, Kanwal, Lal kamal, Padama. *Sans.*: Aravin, Camala, Nalina, Padma, Rajiva, Sara.
Distribution	:	Ponds throughout warmer parts of India.
Parts Used	:	Flowers. Seeds.
Reproductive Cycle	:	Summer–autumn (flowering) and fruiting in winter.
Folk Use	:	Flowers used as adornments for making garlands. Seeds worn as beads.
Active Constituents	:	Flowers possess quercetin, luteolin, isoquercitrin and glucoluteolin.
Trade	:	Indian Lotus.
References	:	Bennet (1986), Birdwood (1880), Chatterjee and Pakrashi (1997), Gogoi and Borthakur (1991), Mehra *et al.* (1975), Singh and Pandey (1998), Sood *et al.* (2005).

Neolitsea chinensis **(Lam.) Chun.**

Syn: *Litsea chinensis* Lamk.; *L. glutinosa* Lour.

Family: Lauraceae

Eng., H. and Sans. Names	:	*Eng.*: Common tallow laurel. *H.*: Garbijaur. *Sans.*: Vasa.
Distribution	:	Throughout India ascending upto 1,350 m in the Himalaya.
Parts Used	:	Leaves. Wood.
Reproductive Cycle	:	May–September.
Folk Uses	:	Leaves tied around the waist and arms for relieving body pain. Wood made into plumes and worn by homicides during their purification ceremony; plumes also worn on belts at dances.
Active Constituents	:	These are naringerin, kaempferol-3 and 7-glucosidin, quercetin and its 3-rhamnoside, cystine, glycine, α and β alanine, valine, tyrosine, proline and leucine (leaves).
References	:	Rastogi and Mehrotra (1970-1979), Radcliffe-Brown(1922), Singh and Kumar (2000), Sood *et al.* (2009).

Nerium indicum **Mill.**

Syn: *N. latifolium* Mill.; *N. oleander* Linn.; *N. ordoratum* Lamk.; *N. ordorum* Solander.; *Thevitia peruviana* (Pers.) Merill.

Family: Apocynaceae

Eng., H. and Sans. Names	:	*Eng.*: Indian oleander, Sweet scented oleander. *H.*: Kaner, Karber, Knel, Kuruvira. *Sans.*: Asvamaraca, Chandata, Karavira, Karvir.
Distribution	:	W. Himalaya.
Part Used	:	Flowers.
Reproductive Cycle	:	Flowering (April-June) and Fruiting in cold season.
Folk Use	:	Flowers used as garlands especially at religious ceremonies.
Active Constituents	:	Flowers yield odoroside A, kaempferol-3-glucoside, a mixture of ursolic andoleanolic acids, β-sitosterol and its glucoside.
Trade	:	Kaner.

References : Jain (1991), Mehra *et al.* (1975), Rastogi and Mehrotra (1970-1979), Sharma (2003), Sood *et al.* (2005), Watt (1889).

Nicotiana tabaccum **Linn.**

Family: Solanaceae

Eng., H. and Sans. Names : *Eng.*: American or Virginian Tobacco.

H.: Bujjer Bhang, Tamaku, Tambaka, Tumak.

Sans.: Dhumrapatrika, Kalanja, Krimighni, Ksharapatra, Tamakhu.

Distribution : Throughout India.

Part Used : Leaves.

Reproductive Cycle : During summer.

Folk Use : Green leaves tied over testes to treat hydrocele.

Active Constituents : Leaves contain 2 isopropyl 5 oxohexanoic acid and D-alanyl-D-alanine.

Trade : Wild Tobacco.

References : Khanna *et al.* (1996), Rastogi and Mehrotra (1960-1969, 1970-1979), Sharma (2003), Sood *et al.* (2005).

Nyctanthes arbortristis **Linn.**

Syn: *Parilium arbor-tristis* Gaertn.; *Scabrita scabra* Linn.; *S. triflora* Linn. Mant.

Family: Oleaceae

Eng., H. and Sans. Names : *Eng.*: Coral jasmine, Indian mourner, Night jasmine, Sorrowful tree.

H.: Harsinghar, Karasli, Kuri, Parja, Saihari, Seoli.

Sans.: Atyuha, Kharapatraka, Parijat, Ragapushpi, Sephalika.

Distribution : Sub-Himalayan region, Chotta Nagpur, Rajasthan, M.P., southwards to Godavari.

Parts Used : Flowers. Stem bark.

Reproductive Cycle : Flowering (July-October) and fruiting in cold season.

Folk Uses : Flowers used for dressing hair and making garlands. Stem bark used for making bracelets and anklets by the tribal people.

Active Constituents : These are myctanthin, tannin, d-manital and glucose(flowers), glycoside and 2 alkaloid (bark).

Trade : Parijata.

References : Chauhan (1999), Gogoi and Borthakur (1991), Jain (1991), Raju (2000), Mehra *et al.* (1975), Sharma (2003), Singh and Pandey (1998), Sood *et al.* (2005).

Nymphaea nouchali **Burm. f.**

Syn: *Castalia pubescens* (Willd.) Blume.; *N. lotus* auct. non Linn.; *N. pubescens* Willd.; *N. rubra* Roxb. *ex* Salisb.

Family: Nymphaeaceae

Eng., H. and Sans. Names : *Eng.*: Indian red water lily.

H.: Alli, Bhenght, Chhota-sundi, Kanval, Koi, Koka.

Sans.: Kahlar, Kamala, Kumuda.

Distribution : Warmer parts of India.

Parts Used : Flowers. Seeds.

Reproductive Cycle : Throughout the year.

Folk Uses : Tribal girls collect the flowers with intact pedicels and split the latter lengthwise into pieces and wear the flowers in their neck like the pendant of a 'mangalsutra'. Seeds employed in making beads for rosaries.

References : Bennet (1986), Singh and Pandey (1998), Sood *et al.* (2005).

Nypa fruticans **Wurmb.**

Family: Arecaceae

Eng. Name : *Eng.*: Nipa palm.

Distribution : Tidal swamps of Sundarbans in Bengal and Andaman Islands.

Part Used : Leaves.

Folk Use : Tender leaves used as head bands mainly by women.

References : Anonymous (1962), Chakrabarty and Balakrishnan (2003).

Ocimum basilicum **Linn.**

Syn: *O. album* Linn.; *O. caryophyllatum* Roxb.; *O. hispidum* Lamk.; *O. minimum* Burm.; *O. pilosum* Willd.

Family: Lamiaceae

Eng., H. and Sans. Names : *Eng.*: Common Basil, Sweet Basil.

H.: Babui tulsi, Gulal tulsi, Kali tulsi, Marua.

Sans.: Munjariki, Surasa, Varvara.

Distribution : Throughout India.

Part Used : Stem.

Reproductive Cycle : June–September.

Folk Use : Stem cut into beads.

References : Ambasta (1986), Dymock *et al.* (1890), Sood and Thakur (2004), Watt (1889).

Ocimum sanctum **Linn.**

Syn: *O. inodorum* Burm.; *O. monachorium* Linn.; *O. tenuifolium* Linn

Family: Lamiaceae

Eng., H. and Sans. Names : *Eng.*: Holy basil, Monk's basil, Rough basil, Sacred basil.

H.: Baranda, Kala tulsi, Tulsi, Varavda.

Sans.: Ajaka, Ban tulsi, Brinda, Divya, Gauri, Hari priya, Laxmi, Manjari, Patrapushpa.

Distribution : Found throughout Himalaya.

Part Used : Roots. Woody Stem. Leaves.

Reproductive Cycle : September–February.

Folk Uses : Roots and woody stem of this perennial herb cut and made into beads. Rosaries composed of its 108 beads extensively used by Vaishnavs a sect of the Hindus for worshipping. Leaves used on garlands.

Active Constituents : Roots contain alkaloids, glycosides, saponins and tannins.

References : Balfour (1871), Dymock *et al.* (1890), Khan and Khanum (2005), Mehra *et al.* (1975), Namboothiry (1980), Sen Gupta (1980), Sood *et al.* (2009).

Opilia amentacea **Roxb.**

Family: Opiliaceae

Distribution : Eastern part of India.

Part Used : Flowers.

Folk Use : Flowers worn by women as adornment.

References : Hosagoudar and Henry (1996a).

Oroxylum indicum **(Linn.) Benth. *ex* Kurz.**

Syn: *Bignonia indica* Linn.; *B. petandra* Lour.;
***Calosanthes indica* Blume.; *Spathodea indica* Pers.**

Family: Bignoniaceae

Eng., H. and Sans. Names : *Eng.*: Indian trumpet tree.

		H.: Arlu, Assarsauna, Sauna, Sonapatha.
		Sans.: Adhvantashatrava, Shyonaka, Svarnavalkana.
Distribution	:	Throughout the greater parts of India mainly in the ravines, moist places in forests.
Parts Used	:	Fruit. Seeds. Bark.
Reproductive Cycle	:	June–January.
Folk Uses	:	Bark used for tattooing. In Kinnaur (H.P.), the seeds are bunched in groups with wire and form a floral type ornament worn on caps. Seeds strung as garlands for idols. Fruit hung at the main door of house or a necklace of seeds around neck prevent contagious diseases from affecting children.
Active Constituents	:	Isolated constituents are baicalein, its 6-glucuronide and 7-glucuronide, scutellarein (stem bark), prunetin, β-sitosterol, glucuronide-oroxidin, synthesis of biacatein-6-glucuronide and 7-glucuronide (seeds).
Trade	:	Arlu, Kinnauri phool.
References	:	Chauhan (1999), Jain and Borthakur (1980), Khan and Khanum (2005), Saklani and Jain (1994), Sen and Behera (2007), Sharma (2003), Sood *et al.* (2009), Watt (1889).

***Oryza sativa* Linn.**

Syn: *O. plena* (Prain.) Chowdhury.

Family: Poaceae

Eng., H. and Sans. Names	:	*Eng.*: Rice, Paddy.
		H.: Chaval, Dhan.
		Sans.: Dhanya, Nivara, Syali, Vrihi.
Distribution	:	Throughout India, cultivated.
Part Used	:	Grains.
Reproductive Cycle	:	May–August.
Folk Use	:	Unhusked grains tied onto bamboo strips to be used as borders for garlands in Thanjaver, Tamil Nadu.
References	:	Ambasta (1986), Francis (1984), Sood *et al.* (2009).

***Osbeckia stellata* Wall.**

Syn: *Melastoma crinata* Roxb.; *O. crinita* Naud.

Family: Melastomataceae

H. Name	:	*H.*: Tsulesi.

Distribution	:	Eastern and N. Himalaya: 1,200-2,400m
Part Used	:	Leaves.
Reproductive Cycle	:	July–October.
Folk Uses	:	Paste of crushed leaves used as a substitute for henna for colouring palm, sole and nails of bride ceremoniously before marriage.
References	:	Bennet (1986), Sood *et al.* (2009).

Oxalis corniculata **Linn.**

Syn: *O. foliosa* Blatt.; *O. pusilla* Salisb.; *O. repens* Thunb.; *O. villosa* Bieb.

Family: Oxalidaceae

Eng., H. and Sans. Names	:	*Eng.*: Indian sorrel, Yellow oxalis. *H.*: Amrul, Anboti, Chalmori, Chuka tripati. *Sans.*: Amlika, Chukrita, Kangeri.
Distribution	:	All regions throughout the warmer parts of India.
Part Used	:	Flowers.
Reproductive Cycle	:	Mostly between September–October.
Folk Use	:	Flowers used as adornments.
Trade	:	Clover sorrel.
References	:	Gogoi and Borthakur (1991), Khan and Khanum (2005), Kirtikar and Basu (1984), Sharma (2003).

Paederia foetida **Linn.**

Syn: *Gentiana scandens* Lour.; *P. scandens* (Lour.) Merr., *P. tomentosa* Blume.

Family: Rubiaceae

Eng., H. and Sans. Names	:	*Eng.*: Chinese flower plant, Kings tonic. *H.*: Bakuchi, Gandhali, Gundali, Khip, Samraji. *Sans.*: Bala, Balya, Charuparni, Gandholi, Prasara, Prasarani.
Distribution	:	Throughout N. E. region.
Part Used	:	Stem.
Reproductive Cycle	:	During rainy season.
Folk Use	:	Pieces of its stem with that of *Tephrosia purpurea* kept inside vagina for inducing abortion.
Active Constituents	:	Important ingredients of stem are hentriacontane, hentriacontanol, methyl mercaptan, ceryl alcohol,

palmitic acid, sitosterol, stigmasterol, campesterol and ursolic acid.

References : Bennet (1986), Jain and Saklani (1992), Khare (2004), Rastogi and Mehrotra (1970-1979), Saklani and Jain (1994).

Pandanus andamanensium Kurz.

Family: Pandanaceae

H. Name : *H.*: Keora.

Distribution : Along sandy and rocky coasts of India.

Part Used : Leaves.

Folk Use : Andaman Island men wear the leaves at the knees, wrists, waist and head; the women make girdles of the leaves.

References : Ambasta (1986), Man (1882).

Pandanus fascicularis Lam.

Syn: *P. amaryllifolius* Roxb.; *P. laevis* Kunth.; *P. latifolius* Hassk.; *P. odoratissimus* Linn.f.; *P. tectorius* Soland. *ex* Parkinson.; *P. variegatus* Miq.

Family: Pandanaceae

Eng., H. and Sans. Names : *Eng.*: Screw pine, Thatch screw pine, Umbrella tree.

H.: Gagandhul, Ketki, Keura, Kewda.

Sans.: Chamarapushpa, Dalapushpa, Halina, Indukalika, Jambuka, Ketaka, Medhya, Naripapriya, Viphala.

Distribution : Coastal regions of Peninsulas, Orissa, U.P., Andaman Islands.

Parts Used : Leaves. Male flowers.

Folk Uses : The Adiyans and Paniyas wear ear-rings known as 'ola' made from its leaves. Leaves also worn on wrist, waist and knees. Male flowers used for hair decoration and as garlands. In W. Coastal regions of India, women commonly use the sweet smelling spathes as well as the staminate flowers as hair ornaments.

Active Constituents : Properties of leaves attributed to β-sitosterol, β-sitostenone, stigmasterol and a new alkaloid–pandamarine.

Trade : Kewda.

References : Dymock (1889a), Jain (1991), Khan and Khanum (2005), Mehra *et al.* (1975), Radhakrishnan *et al.* (2000), Ramachandran *et al.* (1986), Sharma (2003), Singh and Pandey (1998).

Papaver somniferum **Linn.**

Syn: *P. senigerum* DC.

Family: Papaveraceae

Eng., H. and Sans. Names : *Eng.*: Opium poppy, White poppy.

H.: Afim, Afyun, Kashkash, Post.

Sans.: Ahifen, Chosa, Khasa.

Distribution : Native of W. Asia; grows in U.P., Hoshiarpur, Jalandhar, Rajasthan, M.P.

Part Used : Leaves.

Folk Use : Leaves tied on eyes to cure swelling.

References : Jadhav (2006), Prajapati *et al.* (2006)

Passiflora caerulea **Linn.**

Family: Passifloraceae

Distribution : Occur in N. India.

Part Used : Flowers.

Folk Use : Flowers used for hair decoration.

References : Saklani and Jain (1994).

Pavetta tomentosa **Roxb. *ex* J.E. Smith.**

Syn: *Ixora tomentosa* DC.; *P. burnonis* Wight.; *P. canescens* Wall.; *P. crassicaulis* Bremek.; *P. indica* Linn. var. *tomentosa* Roxb.; *P. mollis* Wall.; *P. velutina* Wall.

Family: Rubiaceae

Eng., H. and Sans. Names : *Eng.*: Indian Pellet shrub, White pavetta.

H.: Canchra, Caricara, Kathachampa, Papri.

Sans.: Papata, Pappana.

Distribution : Throughout India.

Part Used : Flowers.

Reproductive Cycle : July–November.

Folk Use : Worn by women as ear rings.

References : Bennet (1986), Hosagoudar and Henry (1996a), Jain (1991), Jain *et al.* (1991), Sood and Sudershna (2008).

Phoenix dactylifera Linn.

Family: Arecaceae

Eng. and H. Names	:	*Eng.*: Date Palm. *H.*: Khajur.
Distribution	:	N. West India.
Part Used	:	Leaves.
Folk Use	:	Chenchus make bangles of the leaves.
References	:	Ambasta (1986), Francis (1984).

Phoenix sylvestris (Linn.) Roxb.

Syn: *Elate sylvestris* Linn.; *Katon indel* Ham.; *P. humilis* Royle.

Family: Arecaceae

Eng., H. and Sans. Names	:	*Eng.*: Date sugar palm, Indian wine palm, The wild date. *H.*: Kejur, Khaji, Khajur, Salma, Sendhi, Thakil. *Sans.*: Kharjura, Khurjjuri, Madhukshir.
Distribution	:	Throughout India.
Parts Used	:	Leaf. Root.
Reproductive Cycle	:	May–June.
Folk Uses	:	To facilitate delivery in Udaipur region of Rajasthan, a small piece of its root kept in the navel of expecting woman during labour pains. 7 of its leaflets together made to touch the whole body to relieve pain due to bee and wasp sting.
References	:	Punjani (2002), Sen and Behera (2007), Singh and Pandey (1998), Sood *et al.* (2005, 2009).

Phragmites karka Trin.

Syn: *Arundo karka* Roxb.; *P. maxima* Blatter and Mc Cann.; *P. roxburghii* (Kunth.) Steud.

Family: Poaceae

Eng., H. and Sans. Names	:	*Eng.*: Nodding reed. *H.*: Nainarakula, Nal, Narkul, Nulanara. *Sans.*: Dhhamana, Nada, Nala, Potagala, Shunya-madhya.
Distribution	:	Marshy places along banks of lakes and streams throughout India ascending upto 1,300m in Himalaya.
Parts Used	:	Inflorescence. Shoot. Culm.

Folk Uses	:	These are kept near the bed of patient to save from any infection. Culm sections worn by Assamese women in ears.
References	:	Anonymous (1962), Kirtikar and Basu (1984), Saklani and Jain (1994), Watt (1889).

***Phyla nodiflora* (Linn.) Greene.**

Syn: *Lippia nodiflora* Mich.; *Verbena nodiflora* Linn.

Family: Verbenaceae

Eng., H. and Sans. Names	:	*Eng.*: Purple lippia. *H.*: Bhujokra, Jalpapli. *Sans.*: Jalapippali, Toyavallari.
Distribution	:	Throughout India upto 900m in wet places.
Part Used	:	Stem.
Folk Use	:	A necklace prepared from small stem pieces hung around the neck to prevent small boils on the body.
References	:	Bennet (1986), Khan and Khanum (2005), Prajapati *et al.* (2006).

***Phyllanthus emblica* Linn.**

Syn: *Cicca emblica* Kurz.; *Dichelastina nodicaulis* Hance.; *Emblica officinalis* Gaertn.; *Phyllanthus glomeratus* Herb.

Family: Euphorbiaceae

Eng., H. and Sans. Names	:	*Eng.*: Emblic myrobalan tree, Gooseberry. *H.*: Amalaci-anola, Amla, Amlaki, Annuli, Anonla. *Sans.*: Amala, Amrita, Amalaki, Adiphala, Akara, Bahuphali, Dhatri, Dhatriphala, Jatiphala, Pancharasa, Shiva, Shriphala, Triphala, Vrishya.
Distribution	:	Tropical India; wild planted from the base of the Himalaya.
Part Used	:	Twig. Fruit.
Reproductive Cycle	:	February–April.
Folk Uses	:	A garland of its twigs around the neck of cattle believed to cure leprosy. Fruit believed to be the model of the 'amalaka' bead, worn by married women.
References	:	Howell (1892), Kirtikar and Basu (1984), Sood *et al.* (2005, 2009).

***Phyllocephalum phyllolaenum* (DC.) Narayana.**

Family: Asteraceae

Distribution : W. Ghats.

Part Used : Stem.

Folk Use : Pieces of its stem with that of *Tricholepis glaberrima* woven together as a necklace for curing throat inflammation.

Reference : Sharma and Singh (2000).

***Pinus kesiya* Royle. *ex* Gordon.**

Syn: *P. kasya* Kurz.; *P. khasya* Royle.; *P. khasyana* Griff.

Family: Pinaceae

Distribution : Khasia mountains, Chittagong: 1,000–2,350m.

Part Used : Pine needles.

Folk Use : Pine needles kept underbneath the pillow for relief against cough and asthma.

References : Ahmed and Borthakur (2005), Watt (1972).

***Pinus wallichiana* A. B. Jackson.**

Syn: *P. chylla* Lodd.; *P. excelsa* Wall. *ex* D. Don.; *P. griffithii* Mc Clell.; *P. pendula* Griff.

Family: Pinaceae

Eng. and H. Names : *Eng.*: Bhutan pine, Five leaved pine, Himalayan blue pine.

H.: Kail.

Distribution : Himalaya: 1,800-4,300m.

Parts Used : Bark. Resin.

Reproductive Cycle : April–June.

Folk Use : Resin and circular bark tied on the organ to keep the bones in place and to facilitate calcification and unification of fractural pieces.

Active Constituents : Turpentine from xylem resin yields α-pinene and isomers of undecane, dodecane and tridecane as minor constituents along with abietic, isopimaric and lambertianic acids.

Trade : Blue pine, Kail.

References : Chauhan (1999), Rastogi and Mehrotra (1985-1989), Sood *et al.* (2009).

***Piper betle* (Linn.) Linn.**

Syn: *Chavica betle* Miq.; *C. siriboa* Linn.; *P. chavya* Ham.; *P. chuvya* Miq.; *P. peepuloides* Wall.; *P. siriboa* Miq.

Family: Piperaceae

Eng., H. and Sans. Names	:	*Eng.*: Betel leaf vine, Betel pepper.
		H.: Pan, Tambuli.
		Sans.: Bhakshyapatra, Bhujangavalli, Divabhishta, Kalaskanda, Nagavalli, Nagini, Parna, Saptalata, Tambula, Vitika.
Distribution	:	Cultivated all over India excepting dry N. Western parts.
Part Used	:	Leaves.
Folk Uses	:	Leaves adorned on the forehead of the bride at the time of marriage. Also warm leaf touched over the knee and belly as a customary ritual before wedding as a test for virginity
Active Constituents	:	Properties of leaves due to 5 propenylphenols chavicol, chavibetol allylpyrocatechol, chavibetol acetate and allyl-pyrocatechol diacetate.
Trade	:	Betel vine.
References	:	Kirtikar and Basu (1984), Rastogi and Mehrotra (1980-1984), Vihari (1995), Watt (1972).

***Pithecellobium angulatum* Benth.**

Syn: *P. clypearia* Benth.; *P. montanum* Benth.; *P. subacutum* Benth.

Family: Fabaceae

Eng. Name	:	*Eng.*: Grasshopper tree.
Distribution	:	E. Himalaya, Sikkim, Khasia, Mishmi and Lushai hills of Assam upto 1,200m, Manipur, Andaman Nicobar Islands.
Part Used	:	Bark.
Folk Use	:	Bark used for tattooing.
References	:	Jain and Borthakur (1980), Saklani and Jain (1994).

***Plumbago indica* Linn.**

Syn: *P. coccinea* Boiss.; *P. rosea* Linn.; *Thela coccinea* Lamk.

Family: Plumbaginaceae

Eng., H. and Sans. Names	:	*Eng.*: Fire plant, Officinal leadwort, Rose-coloured leadwort.

		H.: Chitra, Lalchita, Lalchitarak, Raktacitra.
		Sans.: Davanah, Dipika, Marjara, Pavaka, Raktasikha, Vyala.
Distribution	:	Cultivated throughout India as an ornamental plant; wild in Sikkim and Khasia Hills.
Part Used	:	Root.
Reproductive Cycle	:	During late period of winter.
Folk Uses	:	3-5cm long root piece tied as an amulet on the wrist to alleviate fever and pain. Also 2-3 pieces of its root of 4 cm length worn as necklace with cotton thread for 15-20 days to cure jaundice.
Active Constituents	:	Active properties of roots due to aracidy alcohol, 2-methyl-5-hydroxy-1,4-napthoquinone, β-sitosterol, α-naphthoquinone and α-napthylamine.
References	:	Girach *et al.* (1996), Jain (1991), Khan and Khanum (2005), Pandey *et al.* (1996), Sood and Sudershna (2008).

***Plumbago zeylanica* Linn.**

Syn: *P. auriculata* Blume.; *Thela alba* Lour.

Family: Plumbaginaceae

Eng., H. and Sans. Names	:	*Eng.*: Ceylon lead wort, White leadwort.
		H.: Chita, Chitawar, Chiti, Chitra.
		Sans.: Agnishikha, Analanania, Chitraka, Vaishavan.
Distribution	:	All over India, much cultivated; wild in the W. Peninsula, Bengal, Malay Peninsula.
Part Used	:	Root.
Reproductive Cycle	:	April–September.
Folk Uses	:	Roots tied to leg or hand opposite to the affected part to cure hydrocele. Roots also worn as an amulet around waist for relief from pain of sciatica and on arm in body pain in Bastar.
Active Constituents	:	Characteristic properties of roots due to binaphthaquinone chitranone, zeylinone, elliptinone, droserone, plumbagin and 3-chloroplumbagin,3,3-biplumbagin.
Trade	:	Lead wort.
References	:	Hembrom (1996), Jain (1963, 1991), Khan and Khanum (2005), Sen and Behera (2007), Sinha and Sinha (2001), Sood *et al.* (2005, 2009).

***Pogostemon heyneanus* Benth.**

Syn: *P. patchouli* Hk. f.

Family: Lamiaceae

H. Names	:	*H.*: Pacholi, Peholi.
Parts Used	:	Leaves. Flowers.
Folk Use	:	Leaves and flowers used in garlands.
References	:	Ambasta (1986), Wealth (1969).

***Pongamia pinnata* (Linn.) Pierre.**

Syn: *Dalbergia arborea* Willd.; *Derris indica* (Lamk.) Bennet.; *Galedupa arborea* Roxb.; *G. indica* Lamk.; *P. glabra* Vent.; *Robinia mitis* Linn.

Family: Fabaceae

Eng., H. and Sans. Names	:	*Eng.*: Indian beech, Karang, Pongam oil tree. *H.*: Karaja, Karanj, Karanjaka, Kiramal. *Sans.*: Gita-kaaranja, Karanjah, Naktaahva, Prakiryaa, Udakiryaa.
Distribution	:	Throughout India, very common as a sea-coast tree in the Andamans and Nicobar Islands.
Parts Used	:	Leaves. Root. Pods.
Reproductive Cycle	:	April–June.
Folk Uses	:	Bed of leaves used for curing fever. An amulet of pods on the waist of children considered good for healthy growth and around the neck as a remedy for diptheria and whooping cough. Roots collected on a Saturday and tied to the waist of children to cure syphilis.
Active Constituents	:	These are friedelin, lupeol, lupenone, karanjin, 5-methoxyfurano(4′,5′,8,7) flavone pongapin, 3′methoxypongapin, pongachromene and β-sitosterol (leaves), glabrachrome-II and glabrachalcone (seeds), karanjin (roots).
Trade	:	Indian beech.
References	:	Chauhan (1999), Dwarakan and Ansari (1996), Hosagoudar and Henry (1996a), Rastogi and Mehrotra (1985-1989), Ravi Kumar and Vijaya Sankar (2003), Sen and Behera (2007), Subramaniyam (2000), Tosh (1996).

Porana paniculata Roxb.

Syn: *Dinetus paniculatus* Sweet.

Family: Convolvulaceae

Eng. and H. Names : *Eng.*: Bridal creeper.
H.: Belkamu, Bhowri.

Distribution : Throughout India.

Part Used : Flowers.

Folk Use : Flowers made into garlands for adornment.

References : Singh and Pandey (1998), Sood and Sudershna (2008).

Potentilla leschenaultiana Ser.

Family: Rosaceae

Distribution : Kashmir and Nilgiris: 1,200–4,200 m.

Part Used : Root.

Folk Use : Roots used to make lips red.

Reference : Saklani and Jain (1994).

Pothos scandens Linn.

Syn: *P. angustifolius* (Raf.) C. Presl.; *P. cognatus* Schott.; *P. decipiens* Schott.; *P. longifolius* C. Presl.; *P. roxburghii* De Vriese.

Family: Araceae

Eng. and H. Names : *Eng.*: Climbing pothos, Monkey plant.
H.: Parisa.

Distribution : Throughout India.

Parts Used : Leaves. Stems.

Folk Uses : Leaves used as adornments and stem for making necklaces.

References : Gogoi and Borthakur (1991), Ramachandran *et al.* (1986).

Premna serratifolia Linn.

Syn: *P. corymbosa* auct. non Rottl.; *P. integrifolia* Linn.; *P. obtusifolia* R.Br.; *P. ovalifolia* Wall.; *P. spinosa* Roxb.

Family: Verbenaceae

Eng., H. and Sans. Names : *Eng.*: Head ache tree.
H.: Agetha, Arni, Bhut-bhirravi, Ustabunda.

		Sans.: Agnibijaka, Agnimatha, Ananta, Arani, Araniketu, Ganakasika, Havirmantha, Jayanti, Jyotishka, Kanika, Nadeyi, Pavaka, Tarkari, Vijaya, Vijayantika.
Distribution	:	Andaman coasts. Plains of Assam. Khasia Hills.
Part Used	:	Root.
Reproductive Cycle	:	May–August.
Folk Use	:	Roots collected in the early morning on a Saturday or Sunday from the east of the plant and tied to the waist to cure rheumatism.
Active Constituents	:	Curative principles of roots due to 6 C, β, D gluco-pyranosyl 8 C, β, D xylopyranosyl apigenin.
References	:	Kirtikar and Basu (1984), Rastogi and Mehrotra (1985-1989), Sen and Behera (2007), Sood and Sudershna (2008).

Prunus cerasoides **D. Don**

Syn: *Cerasus puddum* Seringe.; *Maddenia podicellata* Hk. f.; *P. puddum* Roxb. *ex* Wall. *ex* Brandis.

Family: Rosaceae

Eng., H. and Sans. Names	:	*Eng.*: Himalayan wild cherry.
		H.: Padam, Pajia, Phaja.
		Sans.: Padmagandhi, Padmaka, Padmavri, Pita, Pitaka.
Distribution	:	Temperate Himalaya (Garhwal–Sikkim).
Part Used	:	Seeds.
Reproductive Cycle	:	November–May.
Folk Use	:	Tribes of Himalayan hills prepare rosaries and necklaces with the seeds of this large tree.
Active Constituents	:	Stones contain naringenin-4′-methylether-7-xyloside.
Trade	:	Paddam.
References	:	Bennet (1986), Dhiman (2003), Jain (1991), Jain *et al.* (1991), Mehra *et al.* (1975), Raju (2000), Rastogi and Mehrotra (1980-1984), Singh and Rawat (2000).

Prunus mahaleb **Linn.**

Syn: *Cerasus mahaleb* (Linn.) Mill.

Family: Rosaceae

Eng. and H. Names	:	*Eng.*: Mahaleb cherry.
		H.: Gavala.

Distribution	:	Temperate regions of Himalaya, to a lesser extent in the sub–Montane and hilly regions of Peninsular India.
Part Used	:	Kernels.
Reproductive Cycle	:	March–July.
Folk Use	:	Kernels used for making necklaces.
References	:	Ambasta (1986), Guitian (1994), Mehra *et al.* (1975).

Psychotria calocarpa **Kurz.**

Syn: *P. jackii* Hk. f.; *P. viridiflora* Reinw.

Family: Rubiaceae

Distribution	:	Sikkim Himalaya, N. Bengal, Assam, Nagaland, Manipur, Khasia Hills.
Part Used	:	Leaves.
Folk Use	:	Dye obtained from leaves used for tattooing.
References	:	Anonymous (1962), Saklani and Jain (1994).

Pterocarpus marsupium **Roxb.**

Syn: *P. bilobus* Roxb.

Family: Fabaceae

Eng., H. and Sans. Names	:	*Eng.*: Indian kino tree, Malabar kino tree.
		H.: Bija, Bijasal, Bijasar, Pit-shal, Pit-sal.
		Sans.: Araktacandana, Pitasara.
Distribution	:	W. Peninsula, S. India.
Part Used	:	Whole Plant.
Reproductive Cycle	:	Flowers during rainy season and fruiting in summer.
Folk Use	:	Plant used for tattooing on abdomen, arms, back near spinal cord, breast and legs.
Trade	:	Indian Kino.
References	:	Jain (1995), Jain and Sharma (1996), Khan and Khanum (2005).

Pterocarpus santalinus **Linn. f.**

Syn: *P. indicus* Willd.

Family: Fabaceae

Eng., H. and Sans. Names	:	*Eng.*: Red Sanders, Red Sandal Wood.
		H.: Lal-chandan, Raktha-chandan, Rukhto-chandan, Undum.

Sans.: Agaru-gandhakashtaha, Rakta-chandana, Tilapari.

Distribution : Commonly found in South and S. Western India, notably in the states of Karnataka, Chittoor district of A.P. and Chingleput in Tamil Nadu.

Part Used : Wood.

Reproductive Cycle : Flowering (April–May) and fruiting (April–June).

Folk Use : Wood meritorious for making beads for rosaries.

Active Constituents : Contains isoflavone glucosidessavinin, calocedrin and triterpene.

Trade : Red sandal wood.

References : Ambasta (1986), Bennet (1986), Sood *et al.* (2005).

Punica roxburghii **Wall.**

Family: Punicaceae

Part Used : Stones.

Folk Use : Stones stung into necklaces.

Reference : Ramachandran *et al.* (1986).

Quisqualis indica **Linn.**

Syn: *Q. densiflora* Wall. *ex* Miq.

Family: Combretaceae

Eng. and H. Names : *Eng.*: Rangoon creeper.

H.: Rangoon-ki-bel.

Distribution : Cultivated in gardens throughout India.

Part Used : Leaves.

Reproductive Cycle : March–June.

Folk Use : Leaves tied on throat to cure tonsillitis.

References : Ambasta (1986), Jadhav (2006).

Rhynchostylis retusa **(Linn.) Blume**

Syn: *R. garwalia* Reichb. f.; *R. guttata* Reichb.; *R. praemorsa* Blume.; *Saccolabium blumes* Lindl.; *S. garwalicum* Lindl.; *S. guttatum* Lindl.; *S. heathii* Hort.; *S. praemorsum* Lindl.; *S. retusum* Fl.; *S. rheedii* Wight.

Family: Orchidaceae

Eng., H. and Sans. Names : *Eng.*: Foxtail orchid.

H.: Draupadi mala.

Sans.: Gandhanakuli.

Distribution	:	Sikkim and Arunachal Pradesh in N. Eastern India
Part Used	:	Flowers–Inflorescence.
Reproductive Cycle	:	During rainy season in May and June.
Folk Use	:	Flowers or inflorescence used to adorn hairdo of girls during spring festival.
Trade	:	Cat tail orchid.
References	:	Gogoi and Borthakur (1991), Hajra and Baishya (1997), Jain (1991), Sharma (2003), Sood *et al.* (2006).

Ricinus communis Linn.*

Syn: *Croton spinosus* Linn.; *R. inermis* Jacq.; *R. lividus* Jacq.; *R. speciosus* Burm.; *R. spectabilis* Blume.; *R. viridis* Willd.

Family: Euphorbiaceae

Eng., H. and Sans. Names	:	*Eng.*: Castor oil plant, Castor seed, Palma-christi.
		H.: Arand, Arandi, Arend, Erandi, Ind, Rand.
		Sans.: Amanda, Eranda, Ruvuka, Triputihala, Vyadatvaka.
Distribution	:	Largely cultivated on borders of agriculture fields, gardens; wild in wastelands.
Parts Used	:	Leaves. Roots. Seeds.
Reproductive Cycle	:	May–August.
Folk Uses	:	Roots tied to arm as amulet to ward off evil spirits in 'Banswara' district. of Rajasthan. Leaves tied on the forehead twice a day to check headache and for relief from body swelling. Seeds strung as beads.
Active Constituents	:	Leaves contain ricinine, N-demethylricinine, 3-O-β-D-xylopyranosides of kaempferol and quercetin.
Trade	:	Castor bean.
References	:	Bhatt *et al.* (2003), Hembrom (1996), Jain (1994), Rastogi and Mehrotra (1985-1989), Sastry (1966), Sharma (2003), Sood and Sudershna (2008), Sood *et al.* (2005).

Rosa clinophylla Tory.

Syn: *R. involucrata* Roxb. *ex* Lindl.; *R. lyellii* Lindl.; *R. palustris* Buch.

Family: Rosaceae

Eng. Name	:	*Eng.*: The rose vine.
Distribution	:	Westwards to Mt. Abu in Marwar and eastwards to Myanmar along the lower Himalaya from Kumaon eastwards.

Part Used	:	Flowers.
Folk Use	:	Flowers used for hair decoration.
References	:	Bennet (1986), Hooker (1897), Saklani and Jain (1994).

Rosa indica **Linn.**

Syn: *R. chinensis* Jacq.; *R. sinica* Linn.

Family: Rosaceae

Eng. Names	:	*Eng.*: Bengal rose, China rose, Monthly rose.
Distribution	:	Grown in Indian gardens.
Part Used	:	Flowers.
Reproductive Cycle	:	During spring season.
Folk Use	:	Flowers worn as garlands and for decorating hair.
References	:	Ambasta 1986, Auboyer (1961).

Rubia cordifolia **Linn.**

Syn: *R. alata* Wall.; *R. chinensis* Reget.; *R. javana* DC.; *R. manjith* Roxb.; *R. mitis* Miq.; *R. munjista* Roxb.; *R. purpurea* Dcne.; *R. scandens* Zoll.; *R. secunda* Moon.

Family: Rubiaceae

Eng., H. and Sans. Names	:	*Eng.*: The Indian Madder. *H.*: Majith, Manjit, Manjith. *Sans.*: Aruna, Bhandi, Kala, Manjishtha, Manjistha, Meshika.
Distribution	:	Himalaya, Nilgiris and other Hilly areas of India: 1,000-2,500m.
Part Used	:	Root.
Reproductive Cycle	:	June–November.
Folk Use	:	5 cm piece of its root worn as an amulet to relieve headache.
Active Constituents	:	Roots rich in purpurin, munjistin, xantho-purpurein, pseudo-purpurin, alizarin and anthraquinone.
Trade	:	Manjit.
References	:	Aminuddin and Girach (1991), Chauhan (1999), Sinha and Sinha (2001), Sood *et al.* (2005).

Rubus insignis **Hk. f.**

Family: Rosaceae

Distribution	:	E. Tropical Himalaya.
Part Used	:	Spine.
Folk Use	:	Spines used for tattooing.
References	:	Hooker (1897), Saklani and Jain (1994).

Rubus paniculatus **Sm.**

Syn: *R. tiliaceus* Smith.

Family: Rosaceae

Distribution	:	Temperate Himalaya from Rajouri–Sikkim.
Part Used	:	Spine.
Folk Use	:	Spines used for tattooing.
References	:	Hooker (1897), Saklani and Jain (1994).

Rumex scutatus **Linn.**

Family: Polygonaceae

Eng. and H. Names	:	*Eng.*: French sorrel. *H.*: Ambavati
Distribution	:	W. Himalaya.
Part Used	:	Aerial Parts.
Folk Use	:	Aerial parts and those of *Impatiens gigantea* and some lichens grounded into paste with addition of local drink 'sara' for making nail polish.
References	:	Ambasta (1986), Kirtikar and Basu (1984), Sood *et al.* (2001).

Saccharum munja **Roxb.**

Syn: *S. arundinaceum* Retz.; *S. ciliare* T. Anders.; *S. procerum* Roxb.; *S. surpata* Hb. *ex* Wall.

Family: Poaceae

H. Names	:	*H.*: Ekar, Munj, Sara, Sarkanda.
Distribution	:	Throughout the plains and low hills of India.
Part Used	:	Grass.
Reproductive Cycle	:	Flowering in July–August.
Folk Use	:	Girdles of grass used as adornments.
References	:	Ambasta (1986), Shende (1949), Sood *et al.* (2005).

Saccharum officinarum **Linn.**

Syn: *S. sinense* Roxb.

Family: Poaceae

Eng., H. and Sans. Names : *Eng.*: Noble cane, Sugarcane.

H.: Ganna, Ikh, Kumad, Rihu, Uk, Ukh.

Sans.: Adhipatra, Asipatra, Gandidi, Gudada, Gudakashtha, Gudamula, Ikshu, Kantara, Karkotaka, Madhutrina, Pundraka, Rasola, Sastra, Vansha, Vipularasa.

Distribution : Hotter parts throughout India, chiefly in Bihar, Panjab, S. India, U.P.

Parts Used : Leaves. Shoots. Inflorescence.

Reproductive Cycle : January–April.

Folk Use : Used as adornments for dressing hair during dances.

Active Constituents : These are 5 O methylapigenin, 3′,4′,5,7-tetrahydroxy-3,6-dimethoxyflavone (flowers). 5,7-O-dimethylapigenine-4′-O-β-D-glucopyranoside(I) (leaves).

References : Gogoi and Borthakur (1991), Jain (1991), Rastogi and Mehrotra (1970-1979), Sood *et al.* (2009).

Saccharum spontaneum **Linn.**

Syn: *Imperata spontanea* Beauv.; *S. canaliculatum* Roxb.; *S. semidecumbens* Roxb.

Family: Poaceae

Eng., H. and Sans. Names : *Eng.*: Thatch grass, Wild sugarcane.

H.: Kans, Kagara, Kas, Khasi, Kosa, Kuggara.

Sans.: Ikshugandha, Kahsa, Kasalu, Kasha, Khaggara.

Distribution : Throughout warmer parts of India.

Part Used : Inflorescence.

Reproductive Cycle : August–October.

Folk Use : For dressing hair during dances.

References : Dhiman (2003), Jain (1991), Sood and Sudershna, Varghese (1996).

Samadera indica **Gaertn.**

Syn: *S. lucida* (Wall.) Planchon.; *Quassia indica* Noot.

Family: Simarubaceae

Eng., H. and Sans. Names : *Eng.*: Niepa bark tree.

H.: Lokhanti.

Sans.: Gucchakaranjah.

Distribution	:	Evergreen forests and along back waters of S. India.
Part Used	:	Seeds.
Reproductive Cycle	:	Flowering (October–January).
Folk Use	:	Seeds strung into rosaries and tied around children's necks especially for asthma.
References	:	Bennet (1986), Dymock *et al.* (1890), Mehra *et al.* (1975), Watt (1889).

***Sambucus nigra* Linn.**

Family: Caprifoliaceae

Eng. Name	:	*Eng.*: Elder mother.
Distribution	:	Mostly in temperate regions.
Part Used	:	Wood.
Reproductive Cycle	:	Late Spring–early summer.
Folk Use	:	A piece of wood kept by the person believed to provide protection against evil-spirits.
References	:	Rastogi and Mehrotra (1970-1979), Sood *et al.* (2005).

***Santalum album* Linn.**

Syn: *S. myrtifolium* Roxb.; *S. verum* Linn.; *Sirium myrtifolium* Roxb.

Family: Santalaceae

Eng., H. and Sans. Names	:	*Eng.*: Sandal Tree.
		H.: Safed-Chandan, Sandal.
		Sans.: Ananditam, Chandana, Talia-Parnam.
Distribution	:	Karnataka, Tamil Nadu and Kerala, ascending upto 1,200m.
Part Used	:	Wood.
Folk Use	:	Beads formed by cutting wood. Sawdust also pressed into beads at Thanjaver, Tamil Nadu.
Active Constituents	:	α–and β–santalenes, santenol, teresantanol, 1–santonene, santonone, α–santalol and santene.
References	:	Ambasta (1986), Chatterjee and Pakrashi (1997), Dastur (1964), Dymock *et al.* (1890), Prajapati *et al.* (2006).

***Sapindus emarginata* Vahl.**

Syn: *S. abstergens* Roxb.; *S. acutus* Roxb.;
***S. laurifolius* Vahl. var. *emarginatus* (Vahl.) Cooke.; *S. trifoliatus* Linn.**

Family: Sapindaceae

Eng., H. and Sans. Names	:	*Eng.*: The soapnut tree of South India. *H.*: Bara reetha, Ritha. *Sans.*: Arishta phalam, Dhenila, Phenila, Rithakaranja.
Distribution	:	A.P., Karnataka, Maharashtra, Tamil Nadu., Bihar, Orissa, W. Bengal.
Part Used	:	Bark.
Reproductive Cycle	:	December–May.
Folk Use	:	A garland of pieces of its bark around the neck of cattle cures wound of calf-laying.
References	:	Bennet (1986), Jain (1991), Sood *et al.* (2005).

***Saraca asoca* (Roxb.) De Wilde.**

Syn: *Jonesia asoca* Roxb.; *J. pinnata* Willd.; *S. arborescens* Burm.;
***S. indica* Baker.; *S. indica* Linn.; *S. indica* auct. non Linn.;**
***S. minor* Miq.; *S. zollingeriana* Miq.**

Family: Fabaceae

Eng., H. and Sans. Names	:	*Eng.*: Ashoka tree. *H.*: Ashoka. *Sans.*: Ashoka, Kankeli, Vanjula.
Distribution	:	Central and E. Himalaya, E. India, often cultivated in south for its showy flowers.
Part Used	:	Flower.
Reproductive Cycle	:	March–September.
Folk Use	:	Flower garland worn to cure heart palpitation.
Active Constituents	:	Flowers possess β-sitosterol, quercetin, leucocyanidin, kaempferol-3-O-b-D-glucoside, apigenin-7-O-b-D-glucoside and gallic acid.
Trade	:	Asoka.
References	:	Bennet (1986), Jain (1994), Sen and Behera (2007), Sood *et al.* (2005, 2009).

Schima wallichii **(DC.) Korth.**

Syn: *G. chilaunia* Ham.; *Gordonia wallichii* DC.; *S. hypoglauca* Miq.; *S. mollis* Dyer.

Family: Theaceae

Eng. and H. Names	:	*Eng.*: Needle wood.
		H.: Chilauni, Makriya, Makriyachilauni, Makusal.
Distribution	:	E. Himalaya, Khasia Hills, Manipur upto 1,600m.
Part Used	:	Flowers.
Reproductive Cycle	:	April–December.
Folk Use	:	Flowers used for hair decoration in Manipur.
References	:	Bennet (1986), Chatterjee and Pakrashi (1997), Saklani and Jain (1994).

Schrebera sweitenioides **Roxb.**

Family: Oleaceae

Eng., H. and Sans. Names	:	*Eng.*: Weaver's bean tree.
		H.: Banpalas, Gantha, Ghant, Goki, Moka.
		Sans.: Ghanta patali, Mahakapittha, Mokshaka, Musk kaka.
Distribution	:	Sub-Himalayan tract from Kumaon eastwards, C. India.
Part Used	:	Fruit.
Reproductive Cycle	:	Flowers (February-April) and Fruits later.
Folk Use	:	Dried fruits tied to the neck of girls or on the waist of boy to keep evil spirits away.
Active Constituents	:	Fruits contain oleanolic acid and betulinic acid.
References	:	Hosagoudar and Henry (1996a), Kirtikar and Basu (1984), Rastogi and Mehrotra (1980-1984), Sood *et al.* (2005).

Scoparia dulcis **Linn.**

Syn: *Gratiola micrantha* Nutt.; *S. grandiflora* Nash.

Family: Scrophulariaceae

Eng. and H. Names	:	*Eng.*: Sweet bromweed.
		H.: Chini buti, Ghoda tulsi.
Distribution	:	Cultivated fields and roadsides in lowland areas of W. Bengal, Tamil Nadu, S. India.

Part Used	:	Root.
Reproductive Cycle	:	Throughout the year.
Folk Use	:	Root tied on arms for dreams and lactation in Orissa.
Active Constituents	:	A triterpene, mannitol, ifflaionic acid, α-amyrin, betulinic acid, dulcioic acids, friedelin, and coixol are the active principles of roots.
References	:	Banerjee (1974), Jain (1991), Rastogi and Mehrotra (1960-1969, 1980-1984).

Scurrula parasitica **Linn.**

Syn: *Loranthus parasiticus* (Linn.) Merr.; *Taxillus parasiticus* (Linn.) S. T. Chiu.

Family: Loranthaceae

Distribution	:	W. Himalaya of India.
Part Used	:	Whole Plant.
Folk Use	:	Plant if tied to waist causes infertility in women.
Reference	:	Sood *et al.* (2005).

Securinega virosa **Roxb.**

Syn: *Flueggea microcarpa* Blume.; *F. virosa* (Roxb. *ex* Willd.) Baillon.; *S. obvata* Muell. Arg.

Family: Euphorbiaceae

Eng. and H. Names	:	*Eng.*: Small Indian snow berry. *H.*: Bakarcha, Dalme, Patala, Rithoul.
Distribution	:	Throughout India.
Part Used	:	Root.
Reproductive Cycle	:	August–April.
Folk Use	:	A necklace of its roots hung around the neck of the animal cures sprains.
Active Constituents	:	Roots contain norsceurinine and dihydronorsecurinine.
References	:	Bennet (1986), Khan and Khanum (2005), Kumar and Nagiyan (2006), Ramadas *et al.* (2000), Sood and Sudershna (2008).

Selaginella bisulcata **Spreng.**

Family: Selaginellaceae

Distribution	:	N. India.
Parts Used	:	Leaves. Shoots.

Folk Use : These are used as adornments.

Reference : Gogoi and Borthakur (1991).

Selaginella brachystachya **(Hk. and Grev.) Spring.**

Family: Selaginellaceae

Distribution : In Kerala and East Himalaya.

Part Used : Whole plant.

Folk Use : Plant kept over both the ears for inducing sleep.

References : Kumar *et al.* (2000), Sood *et al.* (2005).

Semecarpus anacardium **L. f.**

Syn: *Anacardium latifolium* Lamk.; *A. officinarum* Gaertn.; *S. latifolius* Pers.

Family: Anacardiaceae

Eng., H. and Sans. Names : *Eng.*: Marking nut tree, Oriental cashew.

H.: Bhela, Bhelwa, Bhilawa.

Sans.: Arshagni, Bhallatakab, Bhallika.

Distribution : Throughout sub-Himalayan and tropical regions of India, particularly in deciduous forests of Peninsular India.

Parts Used : Fruit, Seeds.

Reproductive Cycle : May–February.

Folk Uses : A necklace of fruits hungs around the neck of a mother cow for more lactation; also tied to the affected part to cure filaria. Seeds put on babies as amulet in Bastar.

Active Constituents : Fruits yields bhilawanol which is a mixture of cis–and–trans isomers of ursuhenol.

References : Chatterjee and Pakrashi (1997), Elwin (1941), Khan and Khanum (2005), Mukerjee (1944), Rastogi and Mehrotra (1960-1969), Sen and Behera (2007), Sood and Sudershna (2008).

Shorea robusta **Gaertn.**

Syn: *Vatica robusta* Steud.

Family: Dipterocarpaceae

Eng., H. and Sans. Names : *Eng.*: Common sal, Sal Tree.

H.: Sal, Sakhu, Shal.

Sans.: Asvakarna, Guggilam, Rata, Sala.

Distribution	:	Tropical Himalaya. Eastern Districts of C. India. W. Bengal Hills.
Part Used	:	Flowers.
Reproductive Cycle	:	March–July.
Folk Use	:	Hos wear flower in hair.
References	:	Ambasta (1986), Majumdar (1950), Sood *et al.* (2005, 2009).

Shuteria densiflora **Benth.**

Syn: *Shuteria glabrata* Wight. and Arn.; *Glucine vestita* Graham.

Family: Fabaceae

Distribution	:	W. Himalaya.
Part Used	:	Stem.
Folk Use	:	2-3 cm pieces of its stem tied as an amulet on the wrist of a patient suffering from bodyache.
References	:	Girach (2001).

Sida rhombifolia **Linn.**

Syn: *S. canariensis* Willd.; *S. compressa* Wall.; *S. orientalis* Cav.; *S. rhombifolia* Linn. ssp. *retusa* (Linn.) Borssum.; *S. rhombifolia* Linn. var. *retusa* (Linn.) Mast.; *S. retusa* Linn.; *S. rhombifolia* (Linn.) var. *obovata* Wall. *ex mastera.*

Family: Malvaceae

Eng., H. and Sans. Names	:	*Eng.*: Broomjute sida, Paddy's Lucerne, Queen land hemp, Sida hemp. *H.*: Bariara, Bhiunli, Kharenti pitabala, Sahadebi. *Sans.*: Ahikhanda, Atibala, Bala, Barela, Devaarsha, Devbala.
Distribution	:	Common weed throughout tropical and sub-tropical India.
Part Used	:	Root.
Reproductive Cycle	:	July–September.
Folk Use	:	Roots tied to the waist for swelling in testes, and around the abdomen to help in child birth in Assam.
References	:	Bennet (1986), Chauhan (1999), Kirtikar and Basu (1984), Subramaniyam (2000).

Smilax ocreata DC.

Syn: *S. orthoptera* A. DC.; *S. roxburghiana* Wall. *ex* Hk. f.

Family: Liliaceae

Eng. Name	:	*Eng.*: Sarsaparilla.
Parts Used	:	Flowers. Unripe fruits.
Folk Use	:	Used for tattooing.
References	:	Bennet (1986), Jain and Borthakur (1980), Saklani and Jain (1994).

Smilax wightii DC.

Syn: *S. macrophylla* Roxb.; *S. ovalifolia* Roxb.; *S. zeylanica* Wight.

Family: Liliaceae

H. and Sans. Names	:	*H.*: Chob chini, Jangliaushbah, Kumarika, Ramdatun.
		Sans.: Vanamadhusnahi.
Distribution	:	Hilly areas throughout tropical India from Himalaya to Kerala about 2,000m.
Part Used	:	Root.
Reproductive Cycle	:	April-June
Folk Use	:	Pieces of its root kept in bed to check bad dreams.
Active Constituents	:	Diosgenin isolated from roots.
References	:	Jain (1991), Sood *et al.* (2005), Sood and Sudershna (2008), Varghese (1996).

Smithia melenogena var. *incanum* Linn.

Syn: *S. incanum* Linn.

Family: Fabaceae

Part Used	:	Flower Buds.
Folk Use	:	Dye obtained used for painting eyes on auspicious days.
References	:	Jain (1991), Sood *et al.* (2009).

Solanum anguivi Lamk.

Syn: *S. canescens* Blume.; *S. cuneatum* Moench.; *S. heynii* Roem. and Sch.; *S. himalayense* Dunal.; *S. indicum* auct. non Linn.; *S. junghuhnii* Miq.; *S. sodomeum* Linn.; *S. violaceum* Ortega.

Family: Solanaceae

Eng., H. and Sans. Names	:	*Eng.*: Indian night shade, Poison berry.

		H.: Barhanta, Bhatakataiya, Bhirhatta, Kattarha, Vadikadheri.
		Sans.: Brhati, Brihatee.
Distribution	:	Throughout Tropical India.
Part Used	:	Fruit.
Reproductive Cycle	:	Throughout the year.
Folk Use	:	Bhils tie a fruit to the neck of a child suffering from cold and bad throat with the belief that the disease cured as the fruit dries up.
Active Constituents	:	Constituents characterized from fruits are maltase, melibiase, saccharase, diosgenin, lanosterol, β-sitosterol and its glucoside.
Trade	:	Brahati.
References	:	Chatterjee and Pakrashi (1997), Khan and Khanum (2005), Sharma (2003), Singh and Pandey (1998), Sood *et al.* (2005), Watt (1972).

Solanum melongena Linn. var. *incarnum*

Syn: *S. esculentum* Dunal.; *S. ferox* Kurz.; *S. heteracanthum* Dunal.; *S. insanum* Linn.; *S. longum* Roxb.; *S. melanocarpum* Dunal.; *S. ovigerum* Blume.; *S. pressum* Dunal.; *S. sativum* Dunal.; *S. torrum* var. *inerme* Dalz. and Gibs.; *S. trongum* Lamk.; *S. undulatum* Lamk.; *S. zeilanicum* Scop.

Family: Solanaceae

Eng., H. and Sans. Names	:	*Eng.*: Brinjal, Egg plant.
		H.: Badanjan, Baigun, Baingan, Bhanta.
		Sans.: Dirghavartaki, Hingoli, Kantalu, Mahatika, Nilakantaka, Sinhi, Vartakee.
Distribution	:	Widely cultivated for its edible fruits throughout India.
Part Used	:	Flower buds.
Reproductive Cycle	:	June–September.
Folk Use	:	Dye of flower buds used to paint eyes on auspicious days.
References	:	Jain (1991), Sood *et al.* (2009).

Solanum surattense Burm. f.

Syn: *S. armatum* Br.; *S. diffusum* Roxb.; *S. jacquinii* Willd.; *S. maccani* Sant.; *S. virginiaum* Jacq.; *S. xanthocarpum* Schradt.

Family: Solanaceae

Eng., H. and Sans. Names	:	*Eng.*: Dwarf lid brinjal, Wild egg plant, Yellow berried nightshade.

		H.: Katai, Kateli, Ringni, Satyanashi.
		Sans.: Brhati, Kanta, Kari, Nidigadhika, Vyaghri.
Distribution	:	All over India and in Thar Desert.
Part Used	:	Fruit.
Reproductive Cycle	:	April–August.
Folk Use	:	Fruits tied as garland to relieve jaundice and swelling of any part of the body.
Active Constituents	:	Isochlorogenic, neochlorogenic, chlorogenic acids, caffeic acids, solasodine, solasonine and solanocarpine are the important constituents of fruits.
Trade	:	Wild egg plant.
References	:	Bennet (1986), Bhattacharjee (2001), Chauhan (1999), Khan and Khanum (2005), Sharma (2003), Sinha and Sinha (2001).

Spermadictyon sauveolens **Roxb.**

Syn: *Hamiltonia mysorensis* W. and A.; *H. propinqua* Decne.; *H. scabra* Don.; *H. suaveolens* Roxb.

Family: Rubiaceae

Distribution	:	Tropical and Sub-tropical Himalaya.
Part Used	:	Flowers.
Reproductive Cycle	:	July–December.
Folk Use	:	Flowers used for making garlands for adornment both by men and women of Bhil community.
References	:	Bennet (1986), Singh and Pandey (1998).

Sphaeranthus indicus **Linn.**

Syn: *S. africanus* Wall.; *S. hirtus* Willd.; *S. indicus* Willd.; *S. mollis* Roxb.; *S. senegalensis* DC.

Family: Asteraceae

Eng., H. and Sans. Names	:	*Eng.*: East Indian gole thistle.
		H.: Chagul nodi, Gorak mundi, Mundi.
		Sans.: Aruna, Bhikshu, Bhutaghni, Boda, Mahamundi, Mundi, Munditika, Sravani, Vridha.
Distribution	:	Throughout India.
Parts Used	:	Flowers. Stem.
Reproductive Cycle	:	March–June.

Folk Uses	:	A garland of flowers worn around neck to cure small pox and for scenting hair. Tribals hung a wreath made of its stem of 3 cm long in white thread as an amulet on the left arm of a woman after child birth for stopping excessive bleeding.
Active Constituents	:	Flowers characterized by the occurrence of a new eudesmenolide-7α-hydroxyeudesm-4-en-6,12-olide (I) (7-hydroxyfrullanolide).
Trade	:	East Indian Globe Thistle.
References	:	Jain (1963, 1991), Khan and Khanum (2005), Rastogi and Mehrotra (1985-1989), Sharma and Rana (2005), Singh and Narain (2008), Sinha and Sinha (2001), Sood *et al.* (2005).

Spondias pinnata **(L.f.) Kurz.**

Syn: *Evia amara* Comm.; *Mangifera pinnata* L.f.; *S. acuminata* Roxb.; *S. amara* Comm.; *S. macrophylla* Wall. *ex* Hk. f.; *S. mangifera* Willd.

Family: Anacardiaceae

Eng., H. and Sans. Names	:	*Eng.*: Bile tree, Indian hog plum, Wild mango. *H.*: Ambodha, Amra. *Sans.*: Amratakah, Kapichuda, Madhuramlaka, Pitana, Pitanaka, Rasadhya, Tungi, Varshapaki.
Distribution	:	Temperate Himalaya upto 900m.
Part Used	:	Fruit.
Reproductive Cycle	:	April–June.
Folk Use	:	A fruit hung around the neck of children to cure mouth ulcer.
Active Constituents	:	Fruits contain β amyrin, oleanolic acid, glycine, cystine, serine, alanine and leucine.
References	:	Bennet (1986), Khan and Khanum (2005), Rastogi and Mehrotra (1970-1979), Sen and Behera (2007), Sood *et al.* (2009).

Stephania glabra **(Roxb.) Miers.**

Syn: *Cissampelos glabra* (Roxb.) Miers.; *C. finlaysoniansus* Wall.; *C. finlaysoniansus* Wall.; *C. japonica* Miers.; *C. longa* Miers.; *C. roxburghiana* Miers.; *C. roxburhianus* Wall.; *S. rotunda* Hk.; *Clypea wightii* Arn.

Family: Menispermaceae

H. and Sans. Names	:	*H.*: Gindaru. *Sans.*: Nimuka.

Distribution : Hilly districts throughout India.

Part Used : Tuber.

Reproductive Cycle : June–August.

Folk Use : Tuber pieces tied around the neck for swelling of mouth in cattle.

Active Constituents : Curative properties of tubers palmatine, dehydrocorydalmine, palmatrubine stepharanine, tetrahydropalmatine, crydalmine and stepholidine.

References : Rastogi and Mehrotra (1960-1969), Rastogi and Mehrotra (1980-1984), Singh and Kumar (2000), Sood *et al.* (2009).

Stephania hernandifolia (Willd.) Walp.

Syn: *S. japonica* Miers.

Family: Menispermaceae

Eng., H. and Sans. Names : *Eng.*: Tape-vine;

H.: Par.

Sans.: Vanatiktika.

Distribution : Panjab, Dehradun, W. Bengal, Bihar, Assam, E. and W. coasts and S. India.

Part Used : Whole Plant.

Folk Use : Amulet worn on the arm as adornment.

References : Ambasta (1986), Shende (1949).

Stereospermum chelonoides L.f.

Syn: *Bignonia caudata* Miq.; *B. chelonoides* Linn.; *B. colais* Buch-Ham. *ex* Dillw.; *S. colais* (Dillvo.) Mabber.; *S. personatum* (Hassk.) Chatt.; *S. suaveolens* DC.; *S. tetragonum* DC.

Family: Bignoniaceae

Eng., H. and Sans. Names : *Eng.*: Trumpet flower, Yellow snake tree.

H.: Padeer, Padri, Panal.

Sans.: Patali.

Distribution : Throughout moist parts of India.

Parts Used : Fruit. Seeds.

Reproductive Cycle : March–July.

Folk Uses : Cut pieces of mesocarp of fruit strung into garland and its necklace cures migraine, headache and epilepsy.

Pulp of fruits also used for tattooing. A necklace of seeds worn to cure headache.

References : Jamir and Rao (1990), Maliya and Singh (2003), Sen and Behera (2007), Sinha and Sinha (2001).

Streblus asper Lour.

Syn: *Epicarpurus asper* Steud.; *E. orientalis* Bl.; *Trophis aspera* Retz.

Family: Moraceae

Eng., H. and Sans. Names : *Eng.*: Siamese rough bush.

H.: Daheya, Dahia, Karchanua, Rusa, Sahora, Siora.

Sans.: Akshadhara, Bhutavasa, Dhukavasa, Gavakshi, Kshiranasha, Pishachandru, Pitaphala, Rukshapatra, Sakata, Shakhota.

Distribution : Throughout drier regions of India.

Part Used : Twigs.

Reproductive Cycle : March–June.

Folk Uses : Cut pieces of young twigs made into garland and tied around the bitten part as 2-3 bands against snake bite or scorpion sting.

References : Kirtikar and Basu (1984), Kumar and Nagiyan (2006), Maliya (2004), Sood *et al.* (2005).

Striga gesnerioides (Willd.) Vatke.

Syn: *S. orobanchoides* Benth.

Family: Scrophulariaceae

Distribution : W. India, the Deccan from the Concan southwards.

Part Used : Whole Plant.

Folk Use : A reddish purple dye obtained from plant used by tribals to colour the teeth and hands during local festivals.

Reference : Sood *et al.* (2009).

Symphorema polyandrum Wt.

Syn: *S. involucratum* Kew.

Family: Symphoremataceae

Distribution : South Deccan Peninsula in the hills.

Part Used : Fruit.

Folk Use	:	Fruits tied as an amulet on waist to cure rickets among children.
References	:	Girach and Aminuddin (1995), Sood and Sudershna (2008).

Symplocos cochinchinensis (Lour.) Moore.

Syn: *Dicalyx cochinchinensis* Lour.; *Myrtus laruinus* Retz.; *S. ferruginea* Roxb.; *S. laurina* Retz.; *S. loha* Don.; *S. polycarpa* Wall.; *S. racemosa* Wall.; *S. spicata* Roxb.

Family: Symplocaceae

Eng., H. and Sans. Names	:	*Eng.*: Chunga.
		H.: Bholia, Sodh.
		Sans.: Lodra.
Distribution	:	N. E. India.
Part Used	:	Seed.
Reproductive Cycle	:	May–July.
Folk Use	:	Seeds used for making garlands. Seeds also strung into beads and hung around the neck of the children to avert the evil spirits.
References	:	Dhiman (2003), Kumar *et al.* (2000), Sood *et al.* (2009), Watt (1889).

Symplocos racemosa Roxb.

Syn: *S. hamiltoniana* Wall.; *S. nervosa* A.DC.; *S. propinqua* Hance.; *S. rigida* Wall.

Family: Symplocaceae

Eng., H. and Sans. Names	:	*Eng.*: Californian cinchona, China nora, Lodh tree.
		H.: Lodh.
		Sans.: Balabhadra, Balipriya, Bhilli, Hastilodhraka, Shambara, Shukla, Tilaka, Vanarajhata.
Distribution	:	Plains and lower hills of E. and C. India.
Part Used	:	Pollen.
Folk Use	:	Pollen worn by women's on cheek's.
References	:	Dymock *et al.* (1890), Mehra *et al.* (1975).

***Syzygium jambos* Alst.**

Syn: *Eugenia jambos* Linn.; *Jambosa vulgaris* DC.

Family: Myrtaceae

Eng., H. and Sans. Names	:	*Eng.*: Malay apple, Rose apple. *H.*: Gulabjaman. *Sans.*: Brihatphala, Jambu, Jamburaja, Kokileshta, Mahajambu, Nanda.
Distribution	:	Throughout India.
Parts Used	:	Fruit. Flowers.
Reproductive Cycle	:	February–August.
Folk Uses	:	Fruit used as adornments. Flowers worn in N. India.
Active Constituents	:	Alanine, glutamine, threonine, tyrosine, aspartic acid and cysteine isolated from fruits.
References	:	Bennet (1986), Chatterjee and Pakrashi (1997), Gogoi and Borthakur (1991), Kirtikar and Basu (1984), Rastogi and Mehrotra (1960-1969).

***Tabernaemontana divaricata* (Linn.) R. Br. *ex* Roem. *et* Schult.**

Syn: *Ervatamia coronaria* (Jacq.) Stapf.; *E. divaricata* Linn.; *Jasminum zeylanicum* Burm.; *Nerium coronarium* Jacq.; *N. divaricatum* Linn.; *T. coronaria* (Jacq.) Willd.

Family: Apocynaceae

Eng., H. and Sans. Names	:	*Eng.*: East Indian rosebay. *H.*: Chameli, Chandni, Sugandhbala, Tagar. *Sans.*: Nandivriksha, Tagara.
Distribution	:	Ornamental plant in shady places in upper Gangetic Plains, Garwhal, E. Bengal, Khasia Hills, Assam.
Parts Used	:	Leaf, Flowers
Reproductive Cycle	:	July–October.
Folk Uses	:	Leaves tied on boils to suppurate by Jenukurubas. Garlands made from the flowers used by women in Maharashtra, Gujarat and W. Bengal for decorating their coiffures and wearing on the ears.
Active Constituents	:	Flowers contain kaempferol, janetine, hydroxyindolenine and voaphylline whereas leaves yield hyderabadine, apparicine and mehranine.

Trade : Adam's apple.

References : Bennet (1986), Dhiman (2003), Gogoi and Borthakur (1991), Khan and Khanum (2005), Kshirsagar and Singh (2007), Mehra *et al.* (1975), Rastogi and Mehrotra (1980-1984, 1990-1994), Sharma (2003), Singh and Pandey (1998)), Sood *et al.* (2009), Watt (1889).

Tadehagi triquetrum **Ohashi**

Syn: *Desmodium triquetrum* (Linn.) DC.; *Hedysarum triquetrum* Linn.

Family: Fabaceae

Part Used : Flowers.

Reproductive Cycle : Flowering in summer.

Folk Use : Flowers used to make garland.

Reference : Chakrabarty and Balakrishnan (2003).

Tagetes erecta **Linn.***

Syn: *T. major* Gaertn.

Family: Asteraceae

Eng., H. and Sans. Names : *Eng.*: African marigold.

H.: Genda, Gultera, Kalaga, Lalamuraga, Makhamali.

Sans.: Ganduga, Jhandu, Sandu, Sthulapushpa.

Distribution : Throughout India.

Part Used : Flowers.

Reproductive Cycle : August–October.

Folk Use : Used for making garlands, the yellow flowers also worn on the ears for adornment.

Active Constituents : Isolated constituents are mono and diesterified luteins (petals), 6-hydroxykaempferol-7-O-glucoside, quercetagetin and quercetagitrin (flowers) have been isolated.

References : Dhiman (2003), Rastogi and Mehrotra (1970-1979, 1980-1984), Singh and Pandey (1998), Watt (1889).

Tagetes patula **Linn.**

Syn: *T. corymbosa* Sweet.; *T. remotiflora* Kunze.; *T. signata* Bartling.; *T. tenuifolia* Millsp.

Family: Asteraceae

Eng. Name : *Eng.*: French marigold.

Distribution : Grown in Indian gardens upto 1,500m.

Part Used : Flowers.

Folk Use : Used as adornments for making garlands, also worn on the ears.

Active Constituents : Allopatuletin characterized from petals.

References : Gogoi and Borthakur (1991), Rastogi and Mehrotra (1980-1984), Singh and Pandey (1998).

Tamarindus indica **Linn.**

Syn: *T. occidentalis* Gaertn.; *T. officinalis* Hk.

Family: Fabaceae

Eng., H. and Sans. Names : *Eng.*: The tamarind tree.

H.: Ambli, Amli, Anbli, Imli, Nuli, Teter.

Sans.: Ambia, Amlika, Tentrani, Tintidi, Tintiri.

Distribution : Planted throughout India on roadsides and in gardens.

Part Used : Seed.

Reproductive Cycle : June–December.

Folk Use : Seeds made to touch the affected part to relieve pain due to scorpion sting.

Active Constituents : Glucose, galactose, xylose and arabinose isolated from seeds

Trade : Tamarind.

References : Jain (1994), Rastogi and Mehrotra (1980-1984), Sen and Behera (2007), Sood *et al.* (2005).

Tamarix aphylla **(Linn.) Karst.**

Syn: *T. articulata* Vahl.; *T. orientalis* Forsk.; *T. pharas* Ham.; *Thuya aphylla* (Linn.) Karst.

Family: Tamaricaceae

Eng. and H. Names : *Eng.*: Takut galls.

H.: Rakta-jhav.

Distribution : Saline soils of northern belt of India, Rajasthan and Gujarat.

Parts Used : Inflorescence. Wood.

Reproductive Cycle : August–October.

Folk Uses : Inflorescence used for decorating their coiffures during certain social gatherings. Wood cut into beads.

Active Constituents : Inflorescence contains kaempferol-7,4′-dimethylether-3-sulphate and quercetin-3-O-isoferulyl-β-glucuronide.

References : Bennet (1986), Jain (1991), Kumar and Nagiyan (2006), Mehra *et al.* (1975), Rastogi and Mehrotra (1970-1979), Sood *et al.* (2009), Watt (1889).

Tamarix dioica **Roxb.** ***ex*** **Roth.**

Syn: *T. articulata* Wall.

Family: Tamaricaceae

H. and Sans. Names : *H.*: Lal-jhau.

Sans.: Pisula.

Distribution : Throughout N. India along Ganges and near the sea-coast.

Parts Used : Leaves. Shoots.

Folk Use : Used as adornments.

Active Constituents : Leaves possess kaempferide, tamarixetin, D-mannitol, isorhamnetin and trans-2-hydroxy-4-methoxycinnamic acid.

References : Chatterjee and Pakrashi (1997), Gogoi and Borthakur (1991), Rastogi and Mehrotra (1960-1969, 1980-1984).

Taraxacum officinale **Weber.** ***ex*** **Wigg.**

Syn: *Leontodon taraxacum* sensu. Aitchison.; *Taraxacum dens-leonic* Desf.

Family: Asteraceae

Eng. and H. Names : *Eng.*: Bitter wort, Blow ball, Canker time table, Dandelion, Fortune teller, Milk gowan, One O'clock, Peasant's clock, Priests crown.

H.: Baran, Dulal, Kanphul.

Distribution : Throughout the temperate Himalaya.

Part Used : Latex.

Reproductive Cycle : March–November.

Folk Use : Latex of stem used for temporary tattooing on their hands and foreheads by young ladies.

Trade : Dudhli.

References : Chauhan (1999), Sharma (2003), Sood *et al.* (2001).

Taxus baccata Linn.

Syn: *T. contorta* Griff.; *T. nepalensis* Jacq.; *T. nucifer* Wall.; *T. orientalis* Bertoloni.; *T. virgata* Wall.; *T. wallichiana* Zucc.

Family: Taxaceae

Eng., H. and Sans. Names	:	*Eng.*: Common Yew, Yew tree. *H.*: Birmi, Thuno, Zirnub-birmi. *Sans.*: Dhatripatra, Manduparni, Talispatra.
Distribution	:	Temperate Himalaya.
Parts Used	:	Leaves. Twigs. Seeds.
Reproductive Cycle	:	September–November.
Folk Uses	:	Leaves worn as an amulet to prevent conception whereas twigs worn by young unmarried Naga girls as ornaments for decorating their shapely figures. Seeds adorned by Mundas.
Active Constituents	:	Leaves characterized by the occurrence of taxinine, sciadopitysin, ginkgetin, sequoiaflavone, hydrocyanic acid, formic acid, reducing sugars, resins, tannins, ephedrine, ferredoxin, acidyrterone and β-sitosterol.
Trade	:	Himalayan Tree.
References	:	Chauhan (1999), Dhiman (2003), Jain (1991), Mehra *et al.* (1975), Rastogi and Mehrotra (1960-1969), Sood *et al.* (2005, 2009).

Tectona grandis Linn. f.

Family: Verbenaceae

Eng., H. and Sans. Names	:	*Eng.*: Indian oak, Ship tree, The teak tree. *H.*: Sagon, Sagun, Sagwan, Sakhu, Segun. *Sans.*: Anila, Arjunopama, Arna, Atipatraka, Bhumiruha, Dhrubasadhana, Dvaradaru, Gandhasara, Halimaka, Kharachhada, Kolaphala, Mahapatra, Mahiruha, Nau, Saka, Teka, Yogi.
Distribution	:	Western Deccan Peninsula.
Part Used	:	Buds.
Reproductive Cycle	:	August–September.
Folk Use	:	Young apical buds grounded into paste and applied for staining the palms.
References	:	Kirtikar and Basu (1984), Sharma (2000), Sood *et al.* (2005).

Tephrosia purpurea (Linn.) Pers.

Syn: *Cracca purpurea* Linn.; *Galega lanceaefolia* Roxb.; *G. purpurea* Linn.; *T. hamiltonic* Drumm. *ex* Gamble.; *T. lanceolata* R.Grah.; *T. maxima* Linn.; *T. wallichii* Grah. *ex* Fawc. and Rendle.

Family: Fabaceae

Eng., H. and Sans. Names	:	*Eng.*: Purple tephrosia, Wild indigo. *H.*: Dhamasia, Sarpanka, Sarphonka. *Sans.*: Banah, Sharpukha, Vishika-punkhaa.
Distribution	:	Rajasthan and elsewhere in dry places.
Part Used	:	Root.
Reproductive Cycle	:	Flowers during rainy season and Fruiting in winter.
Folk Use	:	Roots tied on the neck of patient suffering from fever.
Active Constituents	:	Roots contain 3′,4′,6,7-tetramethoxyisoflavone.
Trade	:	Sarpankha.
References	:	Bennet (1986), Khan and Khanum (2005), Rastogi and Mehrotra (1960-1969), Sharma (2003), Singh and Pakrashi (1994), Sood and Sudershna (2008).

Terminalia arjuna (Roxb.) Wight. and Arnott.

Syn: *Pentaptera angustifolia* Roxb.; *P. arjuna* Roxb.; *P. glabra* Roxb.; *T. berryi* Wight.; *T. ovalifolia* Rottl

Family: Combretaceae.

Eng., H. and Sans. Names	:	*Eng.*: Arjun, White murdah. *H.*: Anjan, Arjun, Arjuna, Jamla, Kahu, Khawa, Koha. *Sans.*: Arjuna, Dhanvi, Ganduri, Indradru, Kiriti, Kukubha, Partha, Vira.
Distribution	:	All over India.
Part Used	:	Root.
Reproductive Cycle	:	Flowering (April-May) and Fruiting in cold season.
Folk Use	:	Roots collected in the early morning on Saturday or Sunday, tied to the waist to cure intermittent fever.
Trade	:	Arjuna.
References	:	Kirtikar and Basu (1984), Sen and Behera (2007), Sinha and Sinha (2001).

Terminalia bialata **Steud.**

Family: Combretaceae

Eng. Name : *Eng.*: Silvergrey Wood.

Part Used : Leaves.

Folk Use : Leaves used for making garlands.

Trade : White chuglam, Silvergrey.

References : Ambasta (1986), Auboyer (1961).

Terminalia chebula **Retz.**

Syn: *Embryogonia arborea* Teys. and Binn.; *Myrobalanus chebula* Gaertn.; *T. aruta* Ham.; *T. reticulata* Roth.

Family: Combretaceae

Eng., H. and Sans. Names : *Eng.*: Black myrobalan, Gall nut, The chebulic.

H.: Bal-har, Harara, Harra, Har, Pile har, Sanghi har.

Sans.: Abhaya, Haritaki, Pathya.

Distribution : Throughout India.

Part Used : Seed.

Reproductive Cycle : Flowering (April-June) and Fruiting (January-March).

Folk Use : A necklace of seeds worn to prevent small pox.

Active Constituents : Seeds contain palmatic, stearic, oleic, linoleic, arachidic and behenic acids.

Trade : Chebulic myrobalan, Harada, Jagni harad.

References : Jain (1991), Rastogi and Mehrotra (1960-1969), Rastogi and Mehrotra (1970-1979), Sen and Behera (2007), Sinha and Sinha (2001), Sood *et al.* (2005).

Terminalia tomentosa **Wight and Arn.**

Syn: *T. alata* Heyne *ex* Roth.

Family: Combretaceae

Eng., H. and Sans. Names : *Eng.*: Laurel.

H.: Asan, Sain, Saj.

Sans.: Krishanatvaka, Nissaraphalaka, Shyamasa-raka.

Distribution : Most of the parts of India.

Part Used : Leaves.

Folk Use : Leaves worn as the only clothing by Juangs.

References : Ambasta (1986), Elwin (1948).

Thelypteris extensa Bl.

Family: Thelypteridaceae

Distribution	:	N. India.
Part Used	:	Roots.
Folk Use	:	3-4 black roots of the fern put as amulet on the neck of the cattle to cure larval infection.
References	:	Jain and Mudgal (1999), Pal (1997).

Thunbergia fragrans Roxb. var. *laevis.*

Syn: *T. angustifolia* Ham.; *T. javanica* Gaertn.; *T. laevis* Nees.; *T. roxburghia* Nees.; *T. scandens* Pers.; *T. subsagittata* Blanco.

Family: Acanthaceae

Eng. Name	:	*Eng.*: Sweet clock vine.
Distribution	:	Throughout India upto 1,300m.
Parts Used	:	Stem, Fruits.
Folk Uses	:	Fruits tied as necklace to keep off spirits. An amulet of its 7 small pieces of stem on the arm or neck of children worn for 9 days to cure vesicles.
References	:	Bennet (1986), Hosagoudar and Henry (1996b), Sharma and Singh (2000), Sood and Sudershna (2008).

Thysanolaena maxima Roxb.

Syn: *Agrostis maxima* Roxb.; *T. acarifera* Arn. and Nees.; *T. agrostis* Nees.

Family: Poaceae

Eng. and H. Names	:	*Eng.*: Bouquet grass, Broom grass, Tiger grass. *H.*: Phuljanta.
Distribution	:	Throughout India.
Parts Used	:	Blades of grass. Root. Seeds.
Folk Uses	:	Blades of grass kept under the bed to ward off bad dreams. Root of this plant along with that of Prickly chaff flower tied to waist for birth control. Seeds used for adornment among the indigenous communities.
References	:	Jha and Goel (2004), Saklani and Jain (1994), Sood and Sudershna (2008), Sood *et al.* (2005).

***Tinospora cordifolia* (Willd.) Hk.f. and Th.**

Syn: *Cocculus convolvulaceus* DC.; *C. cordifolius* DC.; *C. verrucosus* Wall.; *Menispermum cordifolium* Willd.; *T. palminervis* Miers.

Family: Menispermaceae

Eng., H. and Sans. Names	:	*Eng.*: Gulancha tinospora, Heart leaved moonseed. *H.*: Amrita, Giloya, Gilo, Gurach, Jiwantika. *Sans.*: Amritavalli, Jwaranashini, Madhuparnika, Nirjara.
Distribution	:	Throughout tropical regions of India.
Part Used	:	Stem.
Reproductive Cycle	:	Flowering (March-June) and Fruiting in cold seasons.
Folk Use	:	A necklace of small stem pieces hung around the neck to cure jaundice, fever, general weakness and grey coating of tongue followed by distaste of things.
Active Constituents	:	Active principles of stem due to tinosporidine, β-sitosterol, and a new furanoid diterpene-tinosporide.
Trade	:	Giloy or Gulancha.
References	:	Bhatt *et al.* (2003), Chauhan (1999), Dhiman (2003, 2007), Jain (1994), Kshirsagar and Singh (2007), Punjani (2002), Rastogi and Mehrotra (1970-1979), Sood *et al.* (2005).

***Toddalia asiatica* (Linn.) Lamk.**

Syn: *Paulinia asiatica* Linn.; *Scopolia aculeata* Sm.; *T. aculeata* Pers.; *T. angustifolia* Miq.; *T. floribunda* Wall.; *T. nitida* Lamk.; *T. rubricaulis* Willd.

Family: Rutaceae

Eng., H. and Sans. Names	:	*Eng.*: Forest pepper, Lopez tree, Wild orange tree. *H.*: Dahan, Jangli kalimirch, Kanj. *Sans.*: Dahana, Kaccana, Kanchana, Tiksnaksah.
Distribution	:	Konkan, Deccan, all districts of Tamil Nadu.
Part Used	:	Stem.
Reproductive Cycle	:	February–June.
Folk Use	:	Stem tied around the wrist to check vomiting.
Active Constituents	:	Properties of stem attributed to 2 new coumarins norbraylin and 5, 7, 8-trimethoxycoumarin.
References	:	Bennet (1986), Chatterjee and Pakrashi (1997), Khan and Khanum (2005), Kshirsagar and Singh (2003), Rastogi and Mehrotra (1970-1979), Sood and Sudershna (2008).

Tragopogon dubius **Scop.**

Syn: *T. major* Auct. non Jacq.; *T. pratense* Linn.

Family: Asteraceae

Eng. Names	:	*Eng.*: Bucks-beard, Goat's beard, Meadow-salsify, Yellow salsify.
Distribution	:	W. Himalaya and W. Tibet: 1,500-3,600m.
Part Used	:	Latex.
Folk Use	:	Stem latex used for tattooing by children and young ladies.
Reference	:	Sood *et al.* (2001).

Tribulus terrestris **Linn.**

Syn: *T. lanuginosus* Linn.

Family: Zygophyllaceae

Eng., H. and Sans. Names	:	*Eng.*: Land caltrops, Puncture-vine. *H.*: Bahukantaka, Burragokhur, Chhoto gokhru, Gokhru, Gokshri, Hussuk, Kanta, Kshura. *Sans.*: Gosksura, Laghu gokshura.
Distribution	:	Throughout India upto 3,000m.
Part Used	:	Fruit.
Reproductive Cycle	:	April–August.
Folk Use	:	Fruit after making hole in it attached to the ear lobe opposite to the affected part to relieve hemicrania.
Active Constituents	:	Fruits reportedly contain saponins which produce diosgenin, ruscogenin, gitogenin, kaempferol and quercetin.
Trade	:	Meitcher kund.
References	:	Jain (1994), Khan and Khanum (2005), Punjani (2002), Rastogi and Mehrotra (1985-1989), Sharma (2003), Sinha and Sinha (2001).

Trichodesma indicum **(Linn.) R. Br.**

Syn: *Borago indica* Linn.; *B. spinulosa* Roxb.; *Trichodesma hirsutum* Edgew.; *T. perfoliatum* Wall.; *T. sedgewickianum* Bannerjee

Family: Boraginaceae

H. and Sans. Names	:	*H.*: Andhahuli, Chota-kulpha, Hetamundiya. Ratmundiya. *Sans.*: Adhapushpi, Adhomukha, Avakpushpi, Darvika, Golomi, Romalu, Sarasa, Surasa.

Distribution	:	Open habitats throughout most of India upto about 1,500m. in Himalayas.
Part Used	:	Branches.
Reproductive Cycle	:	July–January.
Folk Use	:	A necklace of branches tied to neck of cattle for curing wounds.
References	:	Patil and Merat (2003), Sood and Sudershna (2008).

Tridax procumbens **Linn.**

Family: Asteraceae

Eng., H. and Sans. Names	:	*Eng.*: Coat buttons, Mexican daisy. *H.*: Khal muriya, Tal muriya, Tikki kasa. *Sans.*: Jayanti.
Distribution	:	Throughout India upto 2,500m.
Part Used	:	Flowers.
Reproductive Cycle	:	Throughout year.
Folk Use	:	Flowers kept in between teeth to reduce toothache.
Active Constituents	:	Curative properties of flowers due to glucoluteolin, luteolin, isoquercetin and quercetin.
References	:	Chatterjee and Pakrashi (1997), Khan and Khanum (2005), Rajendran and Henry (1994), Sood and Sudershna (2008).

Trigonella corniculata **Linn.**

Syn: *T. elatior* Sibth.; *T. esculenta* Willd.; *T. fimbriata* Royle.

Family: Fabaceae

H. and Sans. Names	:	*H.*: Champa methi, Kasuri methi, Kasturi methi, Marwari methi. *Sans.*: Malya.
Distribution	:	W. Himalaya (Kashmir–Bihar), Bengal.
Part Used	:	Flowers.
Reproductive Cycle	:	February–April.
Folk Use	:	Wreaths of flowers adorned by Indian women.
References	:	Ambasta (1986), Dymock (1889a), Kumar and Nagiyan (2006).

***Triticum aestivum* Linn.**

Syn: *T. amyleum* Ser.; *T. compactum* Host Gram.; *T. compositum* Linn.; *T. dicoccum* Schrank.; *T. durum* Desf.; *T. hybernum* Linn.; *T. turgidum* Linn.; *T. vulgare* Villars.

Family: Poaceae

Eng., H. and Sans. Names : *Eng.*: Bread wheat, Common wheat.

H.: Gehun, Giun, Kunak.

Sans.: Arupa, Bahudugdha, Godhuma, Kshiri, Rasala, Sumana, Yava.

Distribution : Cultivated in many parts of India, Deccan Peninsula, Himalaya upto 4,350m.

Part Used : Culms.

Folk Use : Young girls wear ear rings called 'todiya' woven out of culms of this species.

References : Rastogi and Mehrotra (1970-1979), Singh and Pandey (1998), Sood *et al.* (2005).

***Tuber aestivum* Vitt.**

Syn: *T. melanosporum* Vitt.

Family: Tuberaceae

Eng. Name : *Eng.*: The black summer truffle.

Part Used : Tuber.

Folk Use : Amulet of its tuber adorned by people.

References : Malandra (1979), Shende (1949).

***Tylophora indica* (Burm.f.) Merr.**

Syn: *Asclepias asthmatica* L.f.; *A. tunicata* Hort.; A. vomitoria Kaln.; *Cynanchum bracteatum* Thunb.; *C. indicum* Burm.f.; *C. vividiflorum* Sims.; *C. vomitorium* Lamk.; *T.asthmatica* W. and A.; *T. mitroria* Voigt.

Family: Asclepiadaceae

Eng., H. and Sans. Names : *Eng.*: Country ipecacuanha, Emetic swallow, Indian ipecacuanha.

H.: Antamul, Jangli pikvam.

Sans.: Antamuula, Antrapachaka, Gandhana, Latakshiri, Muulini, Shwasaghni.

Distribution : Assam, W. Bengal, Orissa and Peninsular India ascending to 900m.

Part Used : Entire plant.

Reproductive Cycle : July–October.

Folk Use : Worn as a necklace to cure asthma.

Active Constituents : Tylophorine, tylophorinine, a-amyrin, kaempferol and quercetin are the active ingredients of the plant.

Trade : Antamul.

References : Ghosh (2000), Khan and Khanum (2005), Sharma (2003), Sood and Sudershna (2008).

Uraria picta (Jacq.) Desv.

Syn: *Doodia picta* Roxb.; *Hedysarum pictum* Jacq.; *Uraria linearis* Hassk.

Family: Fabaceae

H. and Sans. Names : *H.*: Dabra, Pithvan, Shankaraja, Vishnu jadi.

Sans.: Prishniparni, Prithakparni, Sinhapushpi.

Distribution : Dry grasslands, wasteplaces, open forest in the sub-Himalayan tract (Kashmir–W. Bengal), Assam and all over the plains of India.

Part Used : Stem node of twig.

Reproductive Cycle : July–September.

Folk Use : Stem node of twig worn as a necklace to treat toothache.

Trade : Chitraparni.

References : Chauhan (1999), Sharma (2003), Singh and Narain (2008), Sood and Sudershna (2008).

Urena lobata Linn.

Syn: *U. cana* Wall.; *U. palmata* Roxb.; *U. trilobata* Vell.

Family: Malvaceae

Eng., H. and Sans. Names : *Eng.*: Congujute.

H.: Bachata, Bachita, Brachta, Kunjuya, Lapetua, Pithia, Unga.

Sans.: Atibala, Nagabala, Vanabhenda.

Distribution : N. E. Region.

Part Used : Root.

Reproductive Cycle : July–September.

Folk Use	:	Roots tied to the neck of the cattle alongwith that of *Stephania hernandifolia* (Willd.) for preventing larval/ maggot infection of ulcers.
References	:	Khan and Khanum (2005), Molla and Roy (1996), Sood and Sudershna (2008).

Urtica dioica Linn. (Pl. 8B)

Syn: *U. pilulifera* Linn.

Family: Urticaceae

Eng., H. and Sans. Names	:	*Eng.*: Naughty man's plaything, Nettle, Roman nettle, Stinging nettle, Tinging nettle. *H.*: Bichhu-booti. *Sans.*: Bichhuaa, Chichru.
Distribution	:	N.W. Himalaya from Kashmir–Shimla.
Part Used	:	Roots.
Reproductive Cycle	:	August–October.
Folk Use	:	A necklace of pieces of its root worn on Saturday to ward off any ill effect of black magic.
Active Constituents	:	Homovanillyl alcohol, it's 4'monoglucoside, neoolevil-4-O-β-D-glucoside,9,9'-bisacetyl-neoolevil and its 4-O-β-D-glucoside, sitosterol, 7α and 7β-hydroxy-sitosterols, 6α-diol, vanillin, vanellic acid, and isolariciresinol known to occur in roots.
Trade	:	Stinging nettle.
References	:	Sharma (2003), Sood and Sudershna (2008), Sood *et al.* (2005).

Vanda teres Wall. *ex* Lindl.

Family: Orchidaceae

Distribution	:	Assam, Bengal, Khasia hills, Mayanmar.
Part Used	:	Flowers.
Folk Use	:	Flowers used for personal adornment by womenfolk of Assam.
Reference	:	Sood *et al.* (2006).

***Vanda tessellata* (Roxb.) Hk. f. *ex* Don.**

Syn: *Aerides tessellata* Wight.; *Cymbidium alagnata* Hamilton. *ex* Wall.; *C. tessellatum* Sw.; *C. tesselloides* Roxb.; *Epidendrum tessellatum* Roxb.; *E. tesselloides* Steud.; *V. roxburghii* R.Br.; *V. tesselloides* Reichb.f.

Family: Orchidaceae

H. and Sans. Names	:	*H.*: Banda, Nai, Perasara, Persara, Vanda. *Sans.*: Atirasa, Bhujankhi, Chhatraki, Nakuli, Rasya, Surasa, Vandaka.
Distribution	:	Bengal, Chhota Nagpur, Bihar, M.P., W. Peninsula and Travancore.
Parts Used	:	Leaves, Flowers.
Reproductive Cycle	:	April–June.
Folk Uses	:	Leaves of this orchid are split, strung and worn as anklets by Santhal girls of West Bengal. Flowers used to adorn hair of girls of Assam and Arunachal Pradesh.
Trade	:	Vanda.
References	:	Chauhan (1999), Jain (1991), Mehra *et al.* (1975), Sharma (2003), Sood and Sudershna (2008), Watt (1889).

***Vateria indica* Linn.**

Syn: *V. malabarica* Bl.

Family: Dipterocarpaceae

Eng., H. and Sans. Names	:	*Eng.*: Indian copal, Piney resin, White dammer of south India. *H.*: Chundrus, Kahruba, Safed dammar. *Sans.*: Ajakarnah, Dhupa, Kundura, Pitaphada, Sarjah, Shala.
Distribution	:	Peninsular India from Kanara–Trivandrum.
Part Used	:	Resin Beads.
Reproductive Cycle	:	Flowering (January–April) and fruiting (May–July).
Folk Use	:	Beads from oleoresin resemble amber and worn for decoration.
References	:	Jain (1991), Khare (2004), Mehra *et al.* (1975), Prajapati *et al.* (2006), Watt (1889).

Vernonia cinerea (Linn.) Less.

Syn: *Cocalia rotundifolia* Willd.; *Conyza cenerascenns* Wall.; *C. cinerea* Linn.; *C. heterophylla* Lamk.; *C. incana* Wall.; *C.linifolia* Wall.; *C. mollis* Wild.; *C. ovata* Wall.; *C. prolifera* Lamk.; *Serratula cinerea* Roxb.; *Verinia abbreviata* DC.; *V. albicanns* auct.non DC.; *V. conyzoides* DC.; *V. laxiflora* Less.; *V. leptophylla* DC.; *V. physalifolia* DC.

Family: Asteraceae

Eng., H. and Sans. Names : *Eng.*: Ash coloured fleabane, Purple fleabane.

H.: Dantoapala, Daudopala, Sadodi, Sahadevi.

Sans.: Dandotpala, Devika, Govandani, Saha, Sahadevi, Vishvadeva.

Distribution : Throughout the country upto 2,500m in the Himalaya.

Parts Used : Whole Plant. Root.

Reproductive Cycle : August–May.

Folk Uses : A piece of root tied on the head in the form of cap to get sound sleep. Uprooted whole plant kept under pillow during sleep for insomnia.

Active Constituents : Root contains 26-methyl-heptacosanoic acid, amyrin, campesterol, 24-hydroxytaraxer-14-ene, stigmat-5,17(20)-diene-3β-ol, β-sitosterol, α-spinasterol, stigmasterol, and lupeol acetate.

References : Bhatt *et al.* (2003), Khan and Khanum (2005), Punjani (2002), Sood and Sudershna (2008).

Vetiveria zizanioides (Linn.) Nash.

Syn: *Andropogon muricatus* Hk.; *A. squarrosus* Hk. f. non Linn.f.; *Anatherum zizanioides* (Linn.) Hitch.

Family: Poaceae

Eng., H. and Sans. Names : *Eng.*: Cuscus, Khus-khus, Koosa, Vetiver.

H.: Bala, Bena, Ganrar, Khas-Khas root, Onei, Panni.

Sans.: Abhaya, Amrinala, Bahu-muulaka, Indragupta, Jalavasa, Nalada, Vira.

Distribution : Throughout plains in lower hills of India, mainly on river banks and rich marshy soils.

Part Used : Roots.

Reproductive Cycle : July–October.

Folk Use : An amulet of its root and that of *Pennisetum pedicellatum* tied around neck of pregnant woman to prevent abortion.

Active Constituents	:	Roots possess zizanal, epizizaal, α and β vetivones, khusimone and bicyclodecenone.
Trade	:	Vetiver, Khas.
References	:	Chauhan (1999), Khan and Khanum (2005), Rastogi and Mehrotra (1980–1984), Sharma (2003), Sharma and Singh (2000), Sood and Sudershna (2008).

Vitex negundo Linn.*

Syn: *V. arborea* Des. f. var. *incise* Lamk.; *V. bicolor* Willd.; *V. paniculata* Lamk.; *V. spicata* Lour.; *V. trifolia* Graham. *ex* Chavan. and Oza.; *Nika silvestris* Hermann. *ex* Moldenke

Family: Verbenaceae.

Eng., H. and Sans. Names	:	*Eng.*: Five leaved chaste tree, Indian Privet, Nigandi. *H.*: Mewri, Nengar, Nirgandi, Nisinda, Sambhal, Shawali, Shiwali. *Sans.*: Indrani, Nilapushpa, Nirgundi, Shindhuvara, Shveta-surasa, Surasa, Vrikshaha.
Distribution	:	Throughout India and in warmer zone.
Parts Used	:	Leaves. Fruits. Root.
Reproductive Cycle	:	March–September.
Folk Uses	:	Leaves tied on body to prevent from bad omens. A pillow stuffed with leaves used to get relief from headache and nocturnal emission. Leaves also used as a bed during sleep to relieve rheumatism, sciatica and common cold. Fresh and dried leaves wrapped in a piece of cloth and tied to the waist every day for about one hour to cure waist pain. Root of the plant burnt as incense and then tied on right arm to cure fever. Fruits also tied to opposite side of ear to cure migraine.
Active Constituents	:	These are nishindaside, negundoside (leaves), hentriacontane, β-sitosterol, β-sitosterol acetate and stigmasterol (roots).
Trade	:	Nirgundi-phal, Samhaalu-phal.
References	:	Bhatt *et al.* (2003), Chauhan (1999), Choudhury *et al.* (2008), Kirtikar and Basu (1984), Punjani (2002), Rajendran *et al.* (2003), Sood and Sudershna (2008), Sood *et al.* (2005).

Wattakaka volubilis **(L.f.) Stapf.**

Syn: *Dregea volubilis* (L.f.) Benth. *ex* Hk.f.; *Marsdenia volubilis* Cooke.

Family: Asclepiadaceae

Eng., H. and Sans. Names	:	*Eng.*: Cotton milk plant. *H.*: Ekota, Haran dodi, Nag-chhikni, Nakchiki. *Sans.*: Hemajivanti, Madhumalati, Nakchikni.
Distribution	:	Hotter parts throughout India.
Part Used	:	Leaves.
Reproductive Cycle	:	December–May.
Folk Use	:	Leaves wrapped in cloth used as pillows by persons having sleeplessness.
Active Constituents	:	Leaves contain taraxerol.
References	:	Binu *et al.* (2003), Khan and Khanum (2005), Mehra *et al.* (1975), Sood and Sudershna (2008).

Woodfordia fruticosa **(Linn.) Kurz.**

Syn: *Grislea micropetala* Hocsht *ex* Steud.; *G. punctuata* Ham.; *G. tomentosa* Roxb.; *G. uniflora* Rich.; *Lythrum fruticosum* Linn.; *W. floribunda* Salisb.; *W. tomentosa* Bedd.

Family: Lythraceae.

Eng., H. and Sans. Names	:	*Eng.*: Fire flame bush, Red bell bush, Woodfordia. *H.*: Dhai, Dhaura, Thawi. *Sans.*: Bahupushpika, Dhataki, Rodhrapushpini.
Distribution	:	Throughout greater part of India upto 1,600m in Himalaya.
Part Used	:	Root.
Reproductive Cycle	:	March–June.
Folk Use	:	Roots placed on the bed during parturition for the formation of male child in womb.
Trade	:	Dhai phool.
References	:	Chauhan (1999), Khan and Khanum (2005), Rastogi and Mehrotra (1970-1979, 1990-1994), Sarangi and Sahu (2004), Sood *et al.* (2009).

Wrightia arborea Dennst.

Syn: *Hunteria eugenifolia* Wall.; *Nerium coraia* Wall.; *N. tomentosum* Roxb.; *Periploca arborea* Dennst.; *W. coraia* Wall.; *W. mollissima* Wall.; *W. tomentosa* Wall.; *W. wallichi* A.DC.

Family: Apocynaceae

Eng., H. and Sans. Names	:	*Eng.*: Tomentose wrightia.
		H.: Daira, Dharauli, Dudhi, Dudh–koraiya.
		Sans.: Krishnakutayata.
Distribution	:	Hilly slopes of A.P.
Part Used	:	Bark.
Reproductive Cycle	:	March–July.
Folk Use	:	Bark fibres tied above the snake bitten spot by Jatapus and Savaras.
Active Constituents	:	Isolated constituents from bark are tritriacont-16-one and hexacosan-3,6-diol-12-oic acid.
Trade	:	Dudhi.
References	:	Bennet (1986), Rama Rao and Henry (1996), Rastogi and Mehrotra (1990–1994), Watt (1972).

Wrightia tinctoria (Roxb.) R. Br.

Syn: *Nerium tinctorium* Roxb. var. *rothii.* G. Don.

Family: Apocynaceae

Eng., H. and Sans. Names	:	*Eng.*: Indigo plant, Sweet indrajao.
		H.: Mitha indrajou, Kalikari, Khirni.
		Sans.: Hyamaraka, Kutaja, Sveta kutajah.
Distribution	:	M.P., Rajasthan, W. Peninsula.
Part Used	:	Flowers.
Reproductive Cycle	:	April–August.
Folk Use	:	Flowers made into garlands and worn on ears by Garasias tribe people.
Active Constituents	:	Quercetin and rectin isolated from flowers.
References	:	Dhiman (2003), Rastogi and Mehrotra (1985-1989), Singh and Pandey (1998).

***Xanthium strumarium* Linn.**

Syn: *X. brevirostre* Wallroth.; *X. indicum* DC.; *X. orientale* Blume.; *X. roxburghii* Discolor.

Family: Asteraceae

Eng., H. and Sans. Names	:	*Eng.*: Bur weed, Cocklebur, Clotbur. *H.*: Ban-okra, Chhota-gokhru, Ghagra, Okura. *Sans.*: Arishta, Chanda, Itara, Pitapushpi, Sarpakshi.
Distribution	:	Throughout hotter parts in India.
Parts Used	:	Flowers. Fruit. Seeds.
Reproductive Cycle	:	April–November.
Folk Uses	:	Flowers kept in ear believed to cure headache in Maharashtra. A garland of flowers worn to cure night blindness in Sikkim. Fruits attached to ear lobe opposite to the affected part as ear pendant relieve headache and migrain. Musahar tribes wear a wreath of its fruit in a black thread as a charm against night blindness. Wreath of its seeds worn around child neck on Thursday or Sunday as a treatment for stye.
Active Constituents	:	Fruits comprehend strumaroside, a new polyphenol-1,3,5-tri-O-caffeoylquinic acid along with 3,5-di-O-caffeoylquinic acid.
Trade	:	Adhasis, Burweed.
References	:	Chauhan (1999), Jadhav (2008), Jain (1991), Khan and Khanum (2005), Punjani (2002), Rastogi and Mehrotra (1990-1994), Singh and Narain (2008), Singh and Pandey (1998), Sood *et al.* (2005).

***Xeromphis spinosa* (Thunb.) Keay.**

Syn: *Catunaregam spinosa* (Thunb.) Tiruv.; *Gardienia spinosa* Thunb.; *Randia brandisii* Gamble.; *R. dumetorum* (Retz.) Poir.

Family: Rubiaceae

Eng., H. and Sans. Names	:	*Eng.*: Common emetic nut. *H.*: Aras, Madanphala, Mainphal. *Sans.*: Bastishodhana, Madana, Ramachhardanaka.
Distribution	:	Sub-Himalayan tract, Gujarat, Maharashtra, Tamil Nadu and U.P.
Part Used	:	Fruit.
Folk Use	:	Fruit banded on wrist during marriage ceremony to free ill effects and to remain healthy.

Active Constituents : A new glycoside–randioside A isolated from ripe fruit characterized as β-D-galactopyranosyl(1-3)–oleanolic acid.

References : Bhatt *et al.* (2003), Dhiman (2003), Mehra *et al.* (1975), Rastogi and Mehrotra (1980-1984).

Zanthoxylum limonella **(Dennst.) Alst.**

Syn: *Z. budrunga* Wall. *ex* DC.; *Z. rhetsa* DC.

Family: Rutaceae

Eng., H. and Sans. Names : *Eng.*: Prickly ash.

H.: Badrang.

Sans.: Ashvaghra, Atitejani, Sutejasi.

Distribution : In forests of Konkan, Deccan, N. Kanara, W. Ghats in S. Kanara, Mysore, Malabar, Orissa and Khasia Hills.

Part Used : Wood.

Folk Use : Kacharis use wood for beads and other ornaments.

References : Ambasta (1986), Jain and Dam (1979).

Zizyphus mauritiana **Lamk.**

Syn: *Rhamnus jujube* Linn.; *Z. jujube* Lamk.; *Z. sororia* Schult.; *Z. trinervius* Roth.

Family: Rhamnaceae

Eng., H. and Sans. Names : *Eng.*: Chinese Date, The Indian Jujube.

H.: Baer, Ber, Beri.

Sans.: Badara, Badari, Badri, Kola.

Distribution : Throughout India; wild and extensively cultivated.

Parts Used : Fruits. Wood.

Reproductive Cycle : Flowering(August–September) and fruiting(January–February).

Folk Use : The heart wood and fruits used in making rosaries and beads for worshipping.

References : Sood *et al.* (2005), Varghese (1996).

Zizyphus xylopyrus **Willd.**

Syn: *Z. glaberrima* Santapau.

Family: Rhamnaceae

H. Names : *H.*: Ghont, Gote, Kakor, Kat-ber.

Distribution	:	N.W. India.
Part Used	:	Seeds.
Reproductive Cycle	:	August–January.
Folk Use	:	Kols tie seeds onto wrists of their children.
References	:	Ambasta (1986), Griffiths (1946), Sood *et al.* (2009), Varghese (1996).

Epilogue

The study identified use of 440 species of plants belonging to 347 genera under 121 families by various traditional indigenous Indian societies for adornments, touch therapy and tattooing (Figure 1: Appendix I). Out of them, 1 spp (1 genus: *Caulerpa*) belongs to algae: 3 spp each to fungi (3 genera: *Agaricus, Amanita, Tuber*) and gymnosperms (3 genera: *Gnetum, Pinus, Taxus*); 11 spp (9 genera: *Actinopteris, Adiantum, Asplenium, Equisetum, Helminthostachys, Lygodium, Marsilea, Selaginella, Thelypteris*) to pteridophytes; 78 spp (56 genera: *Acorus, Agave, Allium, Aloe, Alpinia, Anadendrum, Areca, Asparagus, Bambusa, Borassus, Bulbophyllum, Caladium, Calamus, Canna, Caryota, Chlorophytum, Cocos, Coelogyne, Coix, Commelina, Corypha, Crinum, Curculigo, Curcuma,*

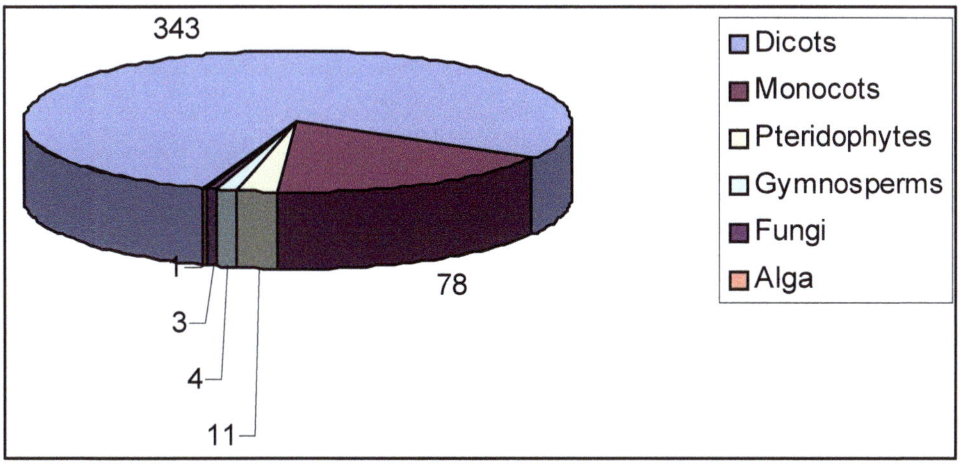

Figure 1: Plants Used for Adornments, Touch Therapy and Tattooing Under Various Divisions of Plant Kingdom

Cynodon, Dendrobium, Dendrocalamus, Desmostachys, Dichrostachys, Didymosperma, Elettaria, Ensete, Globba, Gloriosa, Hordeum, Hornstedtia, Hymenachne, Kaempferia, Licuala, Livistonia, Molineria, Musa, Nannorhops, Nypa, Oryza, Pandanus, Phoenix, Phragmites, Pothos, Rhynchostylis, Saccharum, Smilax, Thysanolaena, Triticum, Vanda, Vetiveria) to monocots; and 343 spp (275 genera) to dicots of angiosperms (Appendix I).

Evidently, the dicotyledonus taxa (78.86 per cent of the total representation) predominate in their use, followed by monocots. Nevertheless, most of these taxa are herbaceous and hold considerable scope for popularization for adornments, touch therapy and tattooing.

Of the estimated diversity in the present database, 275 spp are used for adornments, 189 for touch therapy and 24 for tattooing (Figure 2: Appendix III–V). Three spp such as *Aegle marmelos, Azadirachta indica* and *Oroxylum indicum* are employed commonly for adornments, touch therapy and tattooing (Appendix VI). Relatively, the ratio of spp, genera and families between monocots and dicots is 1: 7.75; 1: 4.91 and 1: 4.39 (Figure 3).

Family-wise analysis indicates that Fabaceae (36 genera, 49 spp), followed by Euphorbiaceae (16 genera, 20 spp), Asteraceae (18 genera, 19 spp), Rubiaceae (14 genera, 15 spp), Verbenaceae (8 genera, 12 spp), Apocynaceae (10 genera, 11 spp), Oleaceae (3 genera, 9 spp), Rutaceae (6 genera, 9 spp), Acanthaceae (6 genera, 8 spp), Combretaceae (4 genera, 7 spp), Cucurbitaceae (6 genera, 7 spp), Malvaceae (6 genera,

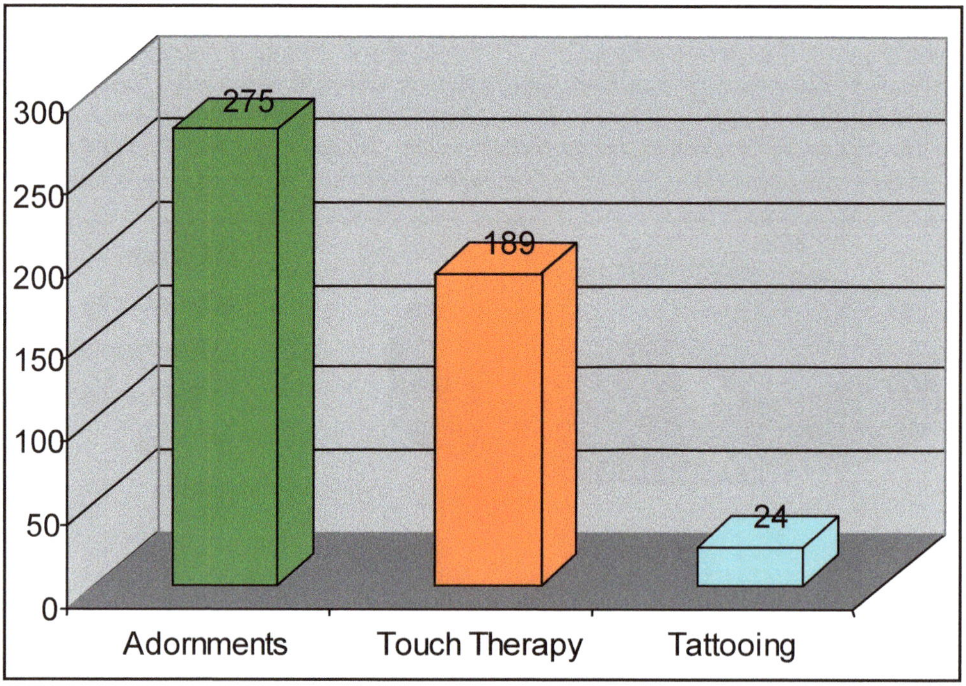

Figure 2: Histogram Showing Total Number of Plants Used for Adornments, Touch Therapy and Tattooing

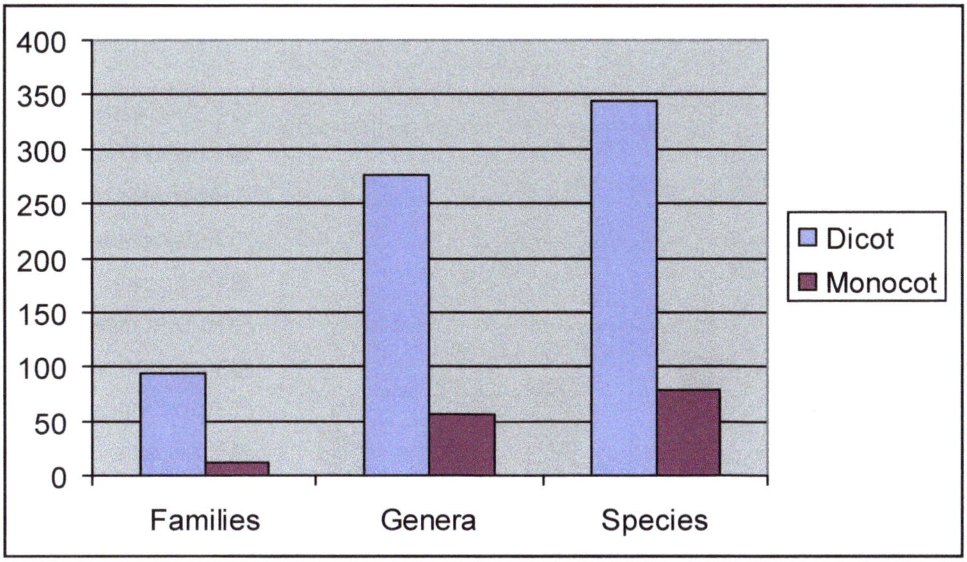

Figure 3: Families, Genera and Species of Adornments, Touch Therapy and Tattooing under Dicots and Monocots

7 spp), Rosaceae (4 genera, 7 spp), Solanaceae (4 genera, 7 spp), Lamiaceae (4 genera, 6 spp), Asclepiadaceae (4 genera, 5 spp), Convolvulaceae (4 genera, 5 spp), Moraceae (2 genera, 5 spp), Anacardiaceae (4 genera, 4 spp), Capparidaceae (3 genera, 4 spp), Elaeocarpaceae (1 genera, 4 spp), Lythraceae (3 genera, 4 spp), Menispermaceae (3 genera, 4 spp) are significant in usages amongst dicots for adornments, touch therapy and tattooing (Figure 5). Amongst monocots, these are: Arecaceae (12 genera, 19 spp), Poaceae (13 genera, 17 spp), Liliaceae (7 genera, 10 spp), Zingiberaceae (6 genera, 8 spp), Orchidaceae (5 genera, 7 spp), Araceae (5 genera, 5 spp), Amaryllidaceae (2 genera, 3 spp), Musaceae (2 genera, 3 spp), Cannaceae (1 genera, 2 spp), Pandanaceae (1 genera, 2 spp) (Figures 4–6; Appendix II, VII). Notably, the size of families has a strong bearing with the above statistics.

Relative percentage distribution of the taxa for predominant Fabaceae amongst dicots is 11.13 and likewise the corresponding figure for Arecaceae amongst monocots is 4.31 (Figure 4; Appendix VII), thereby indicating their overall representation in the present database on adornments, touch therapy and tattooing.

From perusal of information presented in Appendix VIII-IX, it is evident that the most frequently employed genera with a maximum number of spp for adornments are: *Jasminum* (7 spp), *Calamus, Elaeocarpus* (4 spp each), *Acacia, Citrus, Clerodendrum, Coix, Ficus, Saccharum* (3 spp each), *Aeschynomene, Arnebia, Bauhinia, Borassus, Callicarpa, Canna, Caryota, Crinum, Croton, Curcuma, Dendrobium, Erythrina, Euonymus, Hibiscus, Impatiens, Ipomoea, Ixora, Justicia, Livistonia, Murraya, Ocimum, Pandanus, Prunus, Rosa, Symplocos, Tagetes, Tamarix, Terminalia, Vanda, Zizyphus* (2 spp each); and for touch therapy: *Cassia* (3 spp), *Acacia, Allium, Ammannia, Asparagus, Biophytum, Calotropis, Datura, Ficus, Flemingia, Leucas, Lygodium, Melothria, Musa, Pinus, Plumbago, Solanum,*

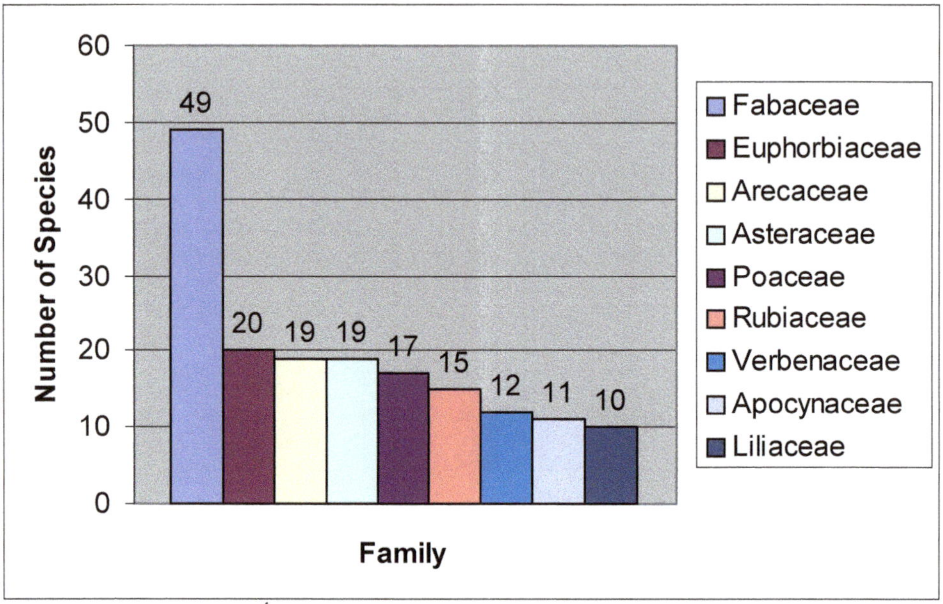

Figure 4: Predominant Families of Plants Used for Adornments, Touch Therapy and Tattooing

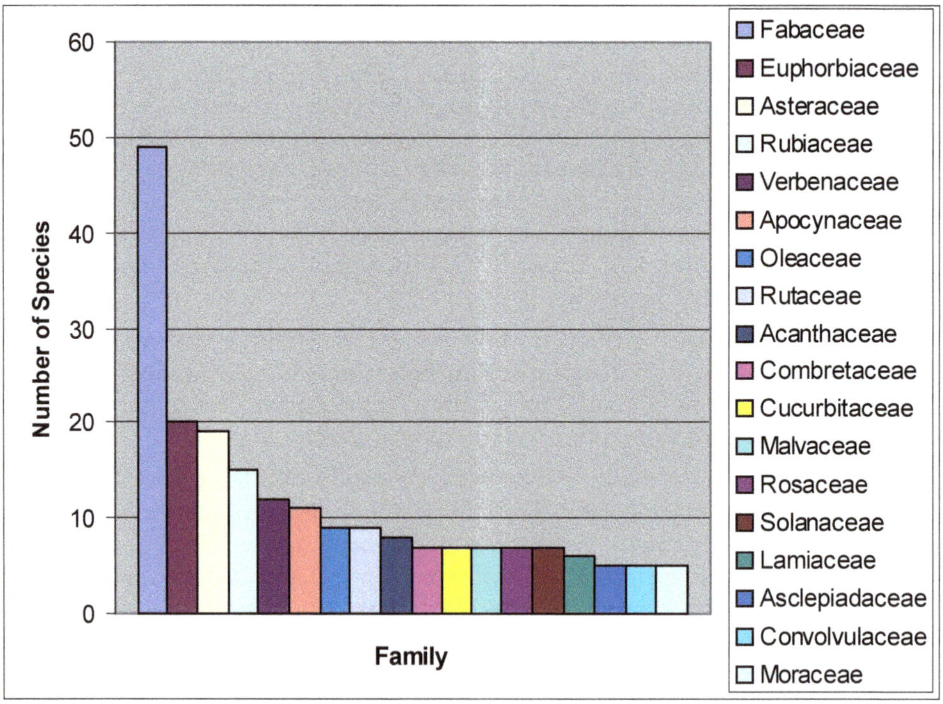

Figure 5: Predominant Dicotyledonous Families Used for Adornments, Touch Therapy and Tattooing

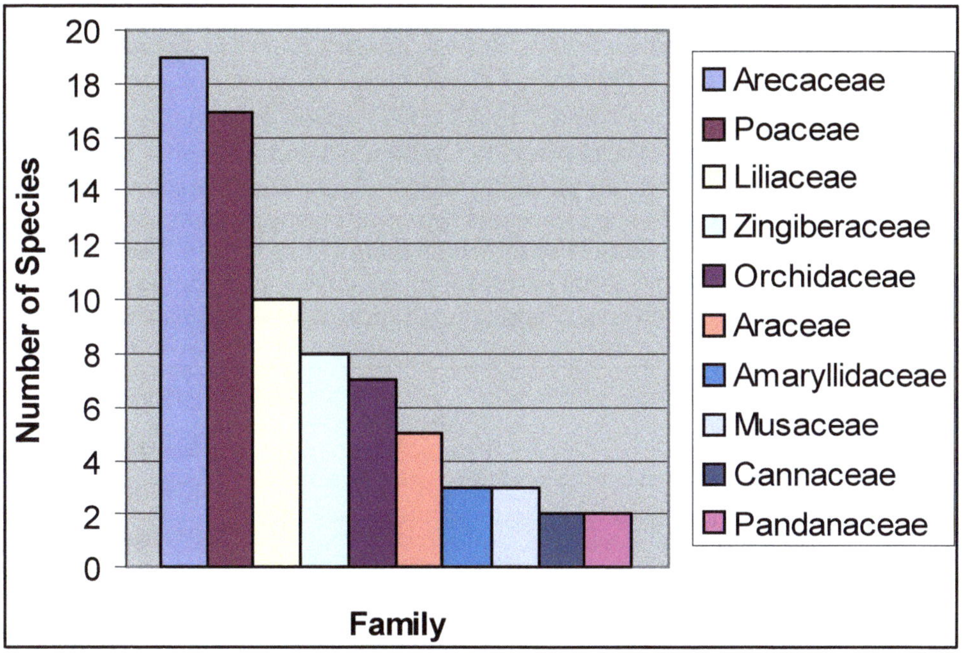

Figure 6: Predominant Monocotyledonous Families Used for Adornments, Touch Therapy and Tattooing

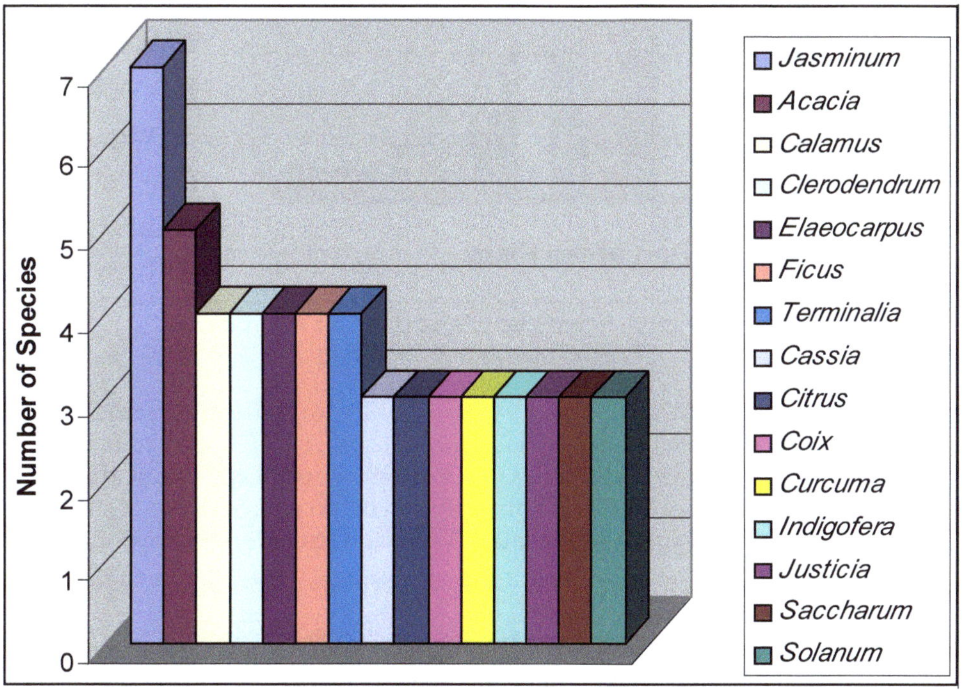

Figure 7: Predominant Genera of Plant Species Used for Adornments, Touch Therapy and Tattooing

Terminalia (2 spp each); and for tattooing: *Rubus* (2 spp), *Aegle, Albizia, Aporusa, Azadirachta, Baccaurea, Baphicacanthus, Berberis, Bombax, Boswellia, Echinops, Eclipta, Euphorbia, Indigofera, Lablab, Oroxylum, Pithecelobium, Psychotria, Pterocarpus, Smilax, Stereospermum, Taraxacum, Tragopogon* (1 spp each) (Figure 7).

Organ-wise analysis of data presented in Figure 8 and Appendix XI revealed that the most predominantly employed usages of plant part/s for adornments are largely met from flowers (94 spp), leaves (44 spp), seeds (39 spp), fruits (23 spp), roots (17 spp), stem (17 spp), wood (15 spp), whole plant (13 spp), stem bark (10 spp), shoots (7 spp), twigs (5 spp), aerial parts (4 spp), beads (4 spp), fibre (4 spp), nuts (4 spp), culms (3 spp), inflorescence (3 spp), pith (3 spp), stones (3spp), flower buds (2

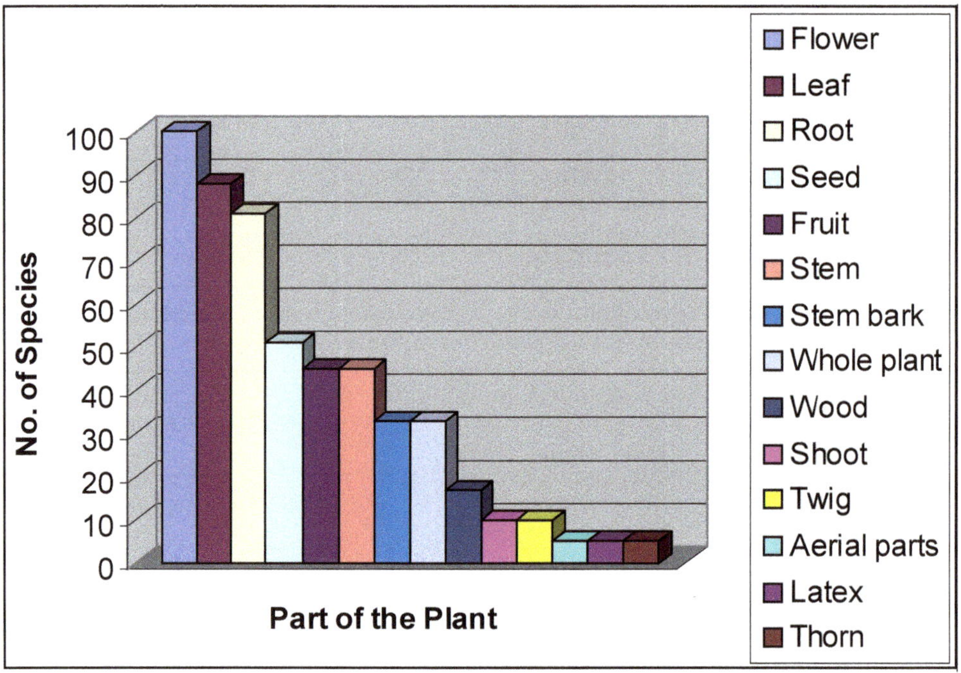

Figure 8: Predominant Plant Parts Used for Adornments, Touch Therapy and Tattooing

spp), kernels (2 spp), and latex (2 spp) (Appendix XII); touch therapy from roots (71 spp), leaves (43 spp), stem (31 spp), whole plant (21 spp), fruit (20 spp), stem bark (16 spp), seeds (14 spp), flowers and twigs (5 spp each), wood (4 spp), bulb, rhizomes, shoots and tubers (2 spp each) (Appendix XIII); and tattooing from those of stem bark (7 spp), leaves (6 spp), thorns (4 spp), latex (3 spp) and fruit (2 spp) (Appendix XIV).

Of the data based on actual field work, use of *Achyranthes aspera, Allium cepa, Asparagus racemosus, Bryophyllum pinnatum, Calotropis procera, Cissampelos pareira, Datura innoxia, Impatiens balsamina, Litsea monopetala, Mirabilis jalapa, Musa paradisiaca, Ricinus communis* and *Vitex negundo* have not been reported earlier and are new records for the field of ethnobotany.

Imperatively, the present work contributes to the database of traditional indigenous knowledge of plants of the country for adornments, touch therapy and tattooing, which have not been documented earlier. Hopefully, the data collected would be utilized for a sustainable management of biodiversity in particular and the society at large.

References

Agarwal, V.S 1986. *Economic Plants of India*. Bishen Singh Mahendra Pal Singh, Dehradun.

Ahmed, A.A. and Borthakur, S.K. 2005. *Ethnobotanical Wisdom of Khasis of Meghalaya*. Bishen Singh Mahendra Pal Singh, Dehradun (India).

Ambasta, S.P. (eds.) 1986. *The Useful Plants of India*. C.S.I.R., New Delhi.

Aminuddin and Girach, R.D. 1991. Ethnobotanical studies on Bondo tribe of district Koraput (Orissa), India. *Ethnobotany* 3: 15-19.

Anonymous, 1948-1976. *Wealth of India*. Vol. I-XI. P.I.D., C.S.I.R., New Delhi.

Anonymous, 1994. *Ethnobiology in India–A Status Report*. Ministry of Environment and Forest, Govt. of India.

Arinathan, V., Mohan, V.R. and John De Britto, A. 2003. Ethno-medicinal survey among Palliyar tribals of Srivilliputhur grizzled giant squirrel wild life sanctuary, Tamil Nadu: 707-714. In Singh, V. and Jain, A.P. (eds.): *Ethnobotany and Medicinal Plants of India and Nepal*. Vol. II. Scientific Publishers, Jodhpur (India).

Asolkar, L.V., Kakkar, K.K. and Chakre, O.J. 1992. *Second Supplement to Glossary of Indian Medicinal Plants with Active Principle*, Part-1 (A-K). C.S.I.R., New Delhi.

Atkinson, E.T. 1882. *Economic Botany of the Himalayan Region*. Cosmo Publ., New Delhi.

Auboyer, J. 1961. *Daily Life in Ancient India*. Wedenfeld and Nicolson, London.

Augustine, J. and Sivadasan, M. 2004. Ethnomedicinal plants of Periyar Tiger Reserve, Kerala, India. *Ethnobotany* 16: 44-49.

Awasthi, A.K. and Goel, A.K. 2000. Ethnobotanical studies on the tribe Onge from Little Andaman Island: 569-575. In Maheshwari, J.K. (eds.): *Ethnobotany and Medicinal Plants of Indian Subcontinent*. Scientific Publishers, Jodhpur.

Balfour, E., (ed.) 1871. *Cyclopedia of India*, Vol. I. Asylum, Scottish, and Foster, Madras.

Banerjee, A. 2000. Ethnobotany of a few plants species in the eroded soil of Birbhum, West Bengal: 527-530. In Maheshwari, J.K. (eds.): *Ethnobotany and Medicinal Plants of Indian Subcontinent*. Scientfic Publishers, Jodhpur.

Banerjee, D.K. 1974. Magico-religious beliefs about plants among some adabasis of India. *Quart. J. Mythic Soc.* 65: 5-6.

Barua, K.N., Barua, I.C. and Das, M. 2000. Ethnobotany of Rajbanshis of Assam: 609-614. In Maheshwari, J.K. (eds.): *Ethnobotany and Medicinal Plants of Indian Subcontinent*. Scientific Publishers, Jodhpur.

Barua, U., Hore, D.K. and Sharma, R. 2007. Wild edible plants of Majuli Island and Darrang districts of Assam. *Indian J. of Trad. Knowledge* 6: 191-194.

Bennet, S.S.R. 1986. *Name Changes in Flowering Plants of India and Adjacent Regions*. Triseas Publication., Dehradun, India.

Bhandary, M.J., Chandrashekhar, K.R. and Kaveriappa, K.M. 1996. Ethnobotany of Gowlis of Uttarakannada District, Karnataka: 244-249. In Maheshwari, J.K. (eds.): *Ethnobotany in South Asia*. Scientific Publishers, Jodhpur.

Bhargava, N. 1997. Plants in folk life and folk lore in Andaman and Nicobar Islands: 225-269. In Jain, S.K. (eds.): *Contribution to Indian Ethnobotany*. Scientific Publishers, Jodhpur.

Bhatt, D.C., Patel, N.K., Mitaliya, K.D. and Ant, H.M. 2003. Herbal magic by contact therapy among tribals and rurals of Gujarat: 205-207. In Singh, V. and Jain, A.P. (eds.): *Ethnobotany and Medicinal Plants of India and Nepal*. Vol I. Scientific Publishers, Jodhpur.

Bhattacharjee, S.K. 2001. *Hand Book of Medicinal Plants*. Pointer Publishers, Jaipur (Raj.) India.

Binu, S., Shanavaskhan, A.E., Santhosh Kumar, E.S. and Pushpangadan, P. 2003. Plants used as Medicine by the Irulas of Palghat district, Kerala, India: 808-814. In Singh, V. and Jain, A.P. (eds.): *Ethnobotany and Medicinal Plants of India and Nepal*. Vol II. Scientific Publishers, Jodhpur.

Birwood, G. 1880. *The Industrial Arts of India*. Chapman and Hall, London.

Borthakur, S.K. 1997. Plants in the folklore and folk life of the Karbis (Mikirs) of Assam: 169-178. In Jain, S.K. (eds.): *Contribution to Indian Ethnobotany*. Scientific Publishers, Jodhpur.

Bushan, J.B. 1964. *Indian Jewellery, Ornaments, and Decorative Designs*. Taraporevala, Bombay.

Chakrabarty, T. and Balarkishnan, N.P. 2003. Ethnobotany of the Andaman and Nicobar Islands India–a review: 869-893. In Singh, V. and Jain, A.P. (eds.):

Ethnobotany and Medicinal Plants of India and Nepal. Vol II. Scientific Publishers, Jodhpur.

Chakraborty, P. 2003. Wild edible plants sold in the daily market in and around of Imphal, Manipur: 481-484. In Singh, V. and Jain, A.P. (eds.): *Ethnobotany and Medicinal Plants of India and Nepal.* Vol. I. Scientific Publishers, Jodhpur.

Chakraborty, M.K. and Bhattacharya, A. 2004. Magico-religious beliefs among the tribal of Purulia district, West Bengal. *J. econ. Taxon. Bot.* 28(2): 510-512.

Chandra, K. 1997. Salai-Pathi (*Licuala peltata* Roxb.–Arecaceae), an essential plant of Jarawa tribe of Andamans. *Ethnobotany* 9: 131-134.

Chandra, M. 1960. Indian costumes and textiles from the eighth to the twelfth century. *J. Indian Textile Hist.* 5: 1-22.

Chandra, M. 1973. *Costumes, Textiles, Cosmetics, and Coiffure in Ancient and Medieval India.* Oriental Publishers, Delhi.

Chandra, S. 1990. *Foundations of Ethnobotany (Pre-1900 Ethnobotany). A Review and Bibliography.* Deep Publications, New Delhi.

Chandra, V. 1997. *Edible Plants of Forestry Origin.* ICFRE, Dehradun.

Chatterjee, A. and Pakrashi, S.C. 1997. *The Treatise on Indian Medicinal Plants.* (Vols. I-V). National Institute of Science Communications (CSIR), New Delhi.

Chattopadhyaya, A. 1976. A note on the importance of *Ficus glomerata* (Udumbara) in ancient Indian society. *J. Indian Hist.* 54: 257-258.

Chaudhari, U.S. and Hutke, V. 2002. Ethno-medico-botanical information on some plants used by Melghat tribes of Amravati district, Maharashtra. *Ethnobotany* 14: 100-102.

Chauhan, N.S. 1999. *Medicinal and Aromatic Plants of Himachal Pradesh.* Indus Publishing Company, New Delhi.

Chopra, R.N., Nayar, S.L. and Chopra, I.C. 1956. *Glossary of Indian Medicinal Plants.* C.S.I.R., New Delhi.

Chopra, R.N., Chopra, I.C. and Varma, B.S.1980. *Supplement to Glossary of Indian Medicinal Plants.* Publications and Information Directorate, New Delhi.

Choudhury, M.D., Shil, S. and Chakraborty, G. 2008. Ethno-medico-botanical studies on Dimasa Kachari of Cachar district, Assam. *Ethnobotany* 20: 128-132.

Chowdhery, H.J. and Wadhwa, B.M. 1984. *Flora of Himachal Pradesh.* BSI, New Delhi.

Cyril Nayagam, M. and Pushparaj, M.S. 2000. 'Touch me not': a medicinal plant of the Nilgiri tribals-a study: 417-420. In Maheshwari, J.K. (eds.): *Ethnobotany and Medicinal Plants of Indian Subcontinent.* Scientific Publishers, Jodhpur.

Dagar, H.S. and Dagar, J.C. 1996. Ethnobotanical studies of the Nicobarese of Chowra island of Nicobar group of Islands: 381-388. In Maheshwari, J.K. (eds.): *Ethnobotany in South Asia.* Scientific Publishers, Jodhpur.

Dagar, H.S. and Dagar, J.C. 2000. Plant folk medicines for gynaecological urino-genital and other related problems among aborigines of Andaman and Nicobar Islands: 561-567. In Maheshwari, J.K. (eds.): *Ethnobotany and Medicinal Plants of Indian Subcontinent*. Scientific Publishers, Jodhpur.

Das, A.K. 1997. Less known uses of plants among the Adis of Arunachal Pradesh. *Ethnobotany* 9: 90-93.

Dhamija, J. 1970. *Indian Folk Arts and Crafts*. National Book Trust, New Delhi.

Dhiman, A.K. 2003. *Sacred Plants and Their Medicinal uses*. Daya Publishers, Delhi.

Dhiman, A.K. 2007. Conservation of some useful medicinal plants of Haridwar district in Uttaranchal State: 147-166. In Chopra, A.K., Khanna, D.R., Prasad, G., Malik, D.S. and Bhutiani, R. (eds.): *Medicinal Plants: Conservation, Cultivation and Utilization*. Daya Publishers, Delhi.

Dastur, J.F. 1964. *Useful Plants of India and Pakistan*. 1977 reprint. Taraporevala, Bombay.

Dikshit, M.G. 1952a. Beads from Ahichchhatra, U.P. *Anci. India* 8: 33-63.

Dikshit, M.G. 1952b. *Some Beads from Kondapur, Hyderabad*. Archaeological Dept. Gov, Hyderabad.

Diwakar, P.G. and Ansari, A.A. 2003. Economic importance of plant resources of Buldhana district of Maharashtra: 592-600. In Singh, V. and Jain, A.P. (eds.): *Ethnobotany and Medicinal Plants of India and Nepal*. Vol II. Scientific Publishers, Jodhpur.

Dutta, M. and Nath, S.C. 2000. Ethno-medico-botany of the Tai-Ahoms of Assam, India: 591-598. In Maheshwari, J.K. (eds.): *Ethnobotany and Medicinal Plants of Indian Subcontinent*. Scientific Publishers, Jodhpur.

Dwarakan, P. and Ansari, A.A. 1996. Less known uses of plants of Kollimalai (Salem district, Tamil Nadu) in South India: 284-286. In Maheshwari, J.K. (eds.): *Ethnobotany in South Asia*. Scientific Publishers, Jodhpur.

Dymock, W. 1889a. The flowers of the Hindu poets. *J. Anthropol. Soc., Bombay* 2: 85-91.

Dymock, W. 1889b. On the use of turmeric in Hindu ceremonial. *J. Anthropol. Soc., Bombay* 2: 441-446.

Dymock, W., Warden, C.J.H. and Hooper, D. 1890. *Pharmacographia Indica*. 1972 reprint. Institute of Health and Tibbi Research, Karachi.

Elwin, V. 1941. The meaning of the cowrie in Bastar. Man in India 21: 198-207.

Ferreira, J.V. 1965. *Totemism in India*. Oxford Univ. Press, London.

Francis, P. 1981. Early human adornment in India, Part I, the Upper Paleolithic. *Bull. Deccan College Postgraduate and Res. Inst.* 40: 137-144.

Francis, P. 1983. Bead report VIII: minor Indian beadmakers. *Ornament* 6: 18-21.

Francis, P. 1984. Plants as Human Adornment in India. *Economic Botany* 38: 194-209.

Francis, P.J. 1997. An interdisciplinary approach to the question of early human adornment in India: 218-231. In Misra, V.D. and Pal, J.N. (eds.) *Indian Prehistory*: 1980. Dept. of Ancient History, Culture and Anthropology. Univ. Allahabad, Allahabad, India.

Furer-Haimendorf, C. von 1933. *The Naked Nagas*. Thaker, Spink, Calcutta.

Furer-Haimendorf, C. von 1943. *The Chenchus: Jungle Folk of the Deccan*. Macmillian, London.

Furer-Haimendorf, C. von 1945. *The Reddis of the Bison Hills*. Macmillian, London.

Gangwar, A.K. and Ramakrishnan, P.S. 1990. Ethnobiological notes on some tribes of Arunachal Pradesh, North Eastern India. *Economic Botany* 44: 94-105.

Ghosh, A. 2000. Herbal veterinary medicine from the tribal areas of Bankura district, West Bengal: 557-560. In Maheshwari, J.K. (eds.): *Ethnobotany and Medicinal Plants of Indian Subcontinent*. Scientific Publishers, Jodhpur.

Girach, R.D. 2001. Ethnobotanical notes on some plants of Mahendragiri hills, Orissa. *Ethnobotany* 14: 52-56.

Girach, R.D. and Aminuddin. 1995. Ethnomedicinal uses of plants among tribals of Singbhum district, Bihar, India. *Ethnobotany* 7: 103-107.

Girach, R.D., Aminuddin, Ahmed, M., Brahmam, M. and Mishra, M.K. 1996. Native phytotherapy among rural population of district Bhadrak, Orissa: 162-164. In Jain, S.K. (eds.): *Ethnobiology In Human Welfare*. Deep Publications, New Delhi.

Gode, P.K. 1938. The antiquity of the Hindu nose-ornament called 'nath'. *Ann. Bhandarkar Res. Inst.* 19: 313-322.

Gogoi, P. and Borthakur, S.K. 1991. Plants in religio-cultural beliefs of the Tai Khamtis of Assam (India). *Ethnobotany* 3: 89-95.

Gopal, G.V. and Jagadish Chandra, K.S. 2003. Ethnobotany of Soligas of Biligirirangana Hills, Karnataka, India: 640-647. In Singh, V. and Jain, A.P. (eds.): *Ethnobotany and Medicinal Plants of India and Nepal*. Vol II. Scientific Publishers, Jodhpur.

Griffiths, W.G. 1946. *The Kol Tribes of Central India*. Royal Asiatic Society, Bengal, Calcutta.

Guitian, J. 1994. Selective fruit abortion in *Prunus mahaleb* (Rosaceae). *Am. Bot.* 81: 1555-1558.

Hajra, P.K. 2004. Plants in medico-religious beliefs in Sanskrit literature, pp 98-103. In Jain S.K., (ed.): *Manual of Ethnobotany*. Scientific Publishers.

Hajra, P.K. and Baishya, A.K. 1997. Ethnobotanical notes on the Miris of Assam Plains: 161-1168. In Jain, S.K. (eds.): *Contribution to Indian Ethnobotany*. Scientific Publishers, Jodhpur.

Hamilton, A., 1955. The people and plant initiative: 10-11. In Martin, G.J. (ed.): *Ethnobotany*. Chapman and Hall, London.

Harborne, J.B. and Baxter, H. 2001. *Chemical Dictionary of Economic Plants.* John Wiley and Sons, Ltd., England.

Hembrom, P.P. 1996. Contact therapy practiced by Mundas of Chotanagpur (Bihar). *Ethnobotany* 8: 36-39.

Hodson, T.C. 1911. *The Naga Tribes of Manipur.* Macmillian, London.

Hooker, J.D. 1872-1897. *The Flora of British India,* Vols. I-VIII. Lalit Mohan Basu, Allahabad.

Hosagoudar, V.B. and Henry, A.N. 1996a. Ethnobotany of Soligas in Biligiri Rangana Betta, Karnataka, Southern India: 228-243. In Maheshwari, J.K. (eds.): *Ethnobotany in South Asia.* Scientific Publishers, Jodhpur.

Hosagoudar, V.B. and Henry, A.N. 1996b. Ethnobotany of tribes Irular, Kurumban and Paniyan of Nilgiris in Tamil Nadu, Southern India: 272-283. In Maheshwari, J.K. (eds.): *Ethnobotany in South Asia.* Scientific Publishers, Jodhpur.

Howell, E.B. 1892. The art industries of the Madras Presidency. *J. Indian Arts and Industries* 40: 29-34.

Hutton, J.H. 1921a. *The Angami Nagas.* 1969 reprint. Oxford, London.

Hutton, J.H. 1921b. *The Sema Nagas.* 1968 reprint. Oxford, London.

Islam, M. 1996. Ethnobotany of certain underground parts of plants of North-Eastern Region, India: 338-343. In Maheshwari, J.K. (eds.): *Ethnobotany in South Asia.* Scientific Publishers, Jodhpur.

Jadhav, D. 2008. Amulets and other plant wearings believed to be contact therapy among tribals of Ratlam district (M.P.) India. *Ethnobotany* 20: 144-146.

Jain, A.K. and Patole, S.N. 2001. Less known medicinal uses of plants among some tribal and rural communities of Pachmarhi Forest (M.P.). *Ethnobotany* 13: 96-100.

Jain, A.K. and Sharma, H.O. 1996. Ethnobiological studies of Sahariya tribe of Central India: 397-399. In Jain, S.K. (eds.): *Ethnobiology in Human Welfare.* Deep Publications, New Delhi.

Jain, S.K. 1963. Magico-religious beliefs about plants among the tribals of Bastar, Madhya Pradesh. *Quart. J. Mythic Soc.* 54: 73-87.

Jain, S.K. 1964. Wild food plants of the tribals of Baster (Madhya Pradesh). *Proc. Nat. Inst. Sci. India* 30 B: 50-80.

Jain, S.K. 1973. (ed.): *Medicinal Plants.* National Book Trust, New Delhi, India.

Jain, S.K. 1981. Observations on ethnobotany of the tribals of Central India: 193-198. In Jain, S.K. (ed.): *Glimpses of Indian Ethnobotany.* Oxford and IBH, Publication, New Dellhi.

Jain, S.K. 1991. *Dictionary of Indian Folk Medicine and Ethnobotany.* Deep Publications, New Delhi.

Jain, S.K. 1994. *Medicinal Plants.* National Book Trust, India.

Jain, S.K. 1995. Ethnobotanical diversity in Zingibers of India. *Ethnobotany* 7: 83-88.

Jain, S.K. 2002. *Bibliography of Indian Ethnobotany*. Scientific Publishers, Jodhpur.

Jain, S.K. 2004. Objective ethnobotany-knowledge traditional, approaches modern. *Ethnobotany* 16: 1-9.

Jain, S.K. and Borthakur, S.K. 1980. Ethnobotany of the Mikirs of India. *Economic Botany* 34: 264-272.

Jain, S.K and Dam, N. 1979. Some ethnobotanical notes from north eastern India. *Economic. Botany*. 33: 52-56.

Jain, S.K. and Mudgal, V. 1999. *A Hand Book of Ethnobotany*. Bishen Singh Mahendra Pal Singh, Dehradun (India).

Jain, S.K., Mudgal, V., Banerjee, D.K., Guha, A., Pal., D.C. and Das. D. 1984. *Bibliography of Ethnobotany*. B.S.I., Howrah.

Jain, S.K. and Saklani, A. 1992. Cross-Cultural Ethnobotanical studies in Northeast India. *Ethnobotany* 4: 25-38.

Jain, S.K. and Sikarwar, R.L.S. 1998. Some Indian plants used in Latin American ethnomedicine. *Ethnobotany* 10: 61-65.

Jain, S.K., Sinha, B.K. and Gupta, R.C. 1991. *Notable Plants in Ethnomedicine of India*. Deep Publications, New Delhi.

Jain, S.K. and Srivastava, S. 1999. *Dictionary of Ethnoveterinary Plants of India*. Deep Publications, New Delhi.

Jain, S.K. and Srivastava, S. 2001. Indian ethnobotanical literature in last two decades– a graphic review and future directions. *Ethnobotany* 13: 1-8.

Jain, S.P. and Singh, S.C. 1997. An ethnO-medicO-botanical survey of Ambikapur district, M.P.: 83-91. In Jain, S.K. (eds.): *Contribution to Indian Ethnobotany*. Scientific Publishers, Jodhpur.

Jamir, N.S. and Rao, R.R. 1990. Fifty new or interesting medicinal plants used by the Zeliangs of Nagaland (India). *Ethnobotany* 2: 11-18.

Jeevan Ram, A. and Venkata Raju, R.R. 2001. Certain potential crude drugs used by tribals of Nallamamais, Andhra Pradesh for skin diseases. *Ethnobotany* 13: 110-115.

Jha, R.R. and Varma, S.K. 1996. Ethnobotany of Sauria Paharias of Santhal Pargana, Bihar: I medicinal plants. *Ethnobotany* 8: 88-91.

Jha, V. and Goel, A.K. 2004. Botanical rosaries and non-flower garlands in folklore life of Mithila region (North Bihar). *Ethnobotany* 16: 27-32.

Kaplesh, I., Rao, V.R., Parabia, F.M., Kothari, I.L. and Parabia, M.H. 2004. Ethnobotany of Little Rann of Kachchh, Gujarat. *Ethnobotany* 16: 40-43.

Katewa, S.S., Choudhary, B.L., Jain, A. and Takhar, H.K. 2001. Some plants in folk medicine of Rajsamand district (Rajasthan). *Ethnobotany* 13: 129-134.

Katewa, S.S. and Sharma, R. 1998. Ethnomedicinal observations from certain Watershed areas of Rajasthan. *Ethnobotany* 10: 46-49.

Khan, I.A. and Khanum, A. 2005. *Role of Biotechnology in Medicinal and Aromatic Plants*. Ukaaz Publications, Hyderabad, Andhra Pradesh.

Khanna, K.K. 2002. Unreported ethnomedicinal uses of plants from the tribal and rural folklores of Gonda district, Uttar Pradesh. *Ethnobotany* 14: 52-56.

Khanna, K.K., Mudgal, V., Shukla, G. and Srivastava, P.K. 1996. Unreported Ethnomedicinal Uses of Plants from Mirzapur district, Uttar Pradesh: 112-117. In Maheshwari, J.K. (eds.): *Ethnobotany in South Asia*. Scientific Publishers, Jodhpur.

Khare, C.P. 2004. *Encyclopedia of Indian Medicinal Plants*. Springer-Verlag, Berlin Heidelberg.

Khare, C.P. 2007. *Indian Medicinal Plants*. Springer-Verlag, Berlin Heidelberg.

Khumbongmayum, A.D., Khan, M.L. and Tripathi, R.S. 2005. Ethnomedicinal plant in the sacred groves of Manipur. *Indian J. Trad. Knowledge* 4(1): 21-32.

Kirtikar, K.R. and Basu, B.D. 1984. *Indian Medicinal Plants*. Vols. I-IV. Lalit Mohan Basu, Allahabad, India.

Kosambi, D.D. 1962. *Myth and Reality*. Popular Press, Bombay.

Krishnamurthy, S. 1953. *Horticultural and Economical Plants of the Nilgiris*. Government Botanic Gardens Cenetenary Committee, Udhagamandalam.

Kshirsagar, R.D. and Singh, N.P. 2000. Less known ethnomedicinal uses of plants in Coorg district of Karnataka state, Southern India. *Ethnobotany* 12: 12-16.

Kshirsagar, R.D. and Singh, N.P. 2003. Less known herbal remedies from Soligas of Mysore, Southern India: 656-661. In Singh, V. and Jain, A.P. (eds.): *Ethnobotany and Medicinal Plants of India and Nepal*. Vol. II. Scientific Publishers, Jodhpur.

Kshirsagar, R.D. and Singh, N.P. 2007. *Ethnobotany of Mysore and Coorg, Karnataka State*. Bishen Singh Mahendra Pal Singh, Dehradun (India).

Kumar, K.K., Sasidharan, N. and Swarupanandan, K. 2000. Ethnobotanical studies on the hill tribes in the Shola forests of high ranges Kerala, South India: 451-466. In Maheshwari, J.K. (eds.): *Ethnobotany and Medicinal Plants of Indian Subcontinent*. Scientific Publishers, Jodhpur.

Kumar, R.V. and Pullaiah, T. 1998. Medicinal plants used by the tribals of Prakasam district, Andhra Pradesh. *Ethnobotany* 10: 97-102.

Kumar, S. 1999. *The Economic Plants of North-East India*. Scientific Publishers, Jodhpur.

Kumar, S. 2002. *The Medicinal Plants of North East India*. Scientific Publishers, Jodhpur.

Kumar, S., Goyal, S. and Parveen, F. 2004. Ethnoveterinary plants in Indian Arid Zone. *Ethnobotany* 16: 91-95.

Kumar,S. and Nagiyan, P. 2006. Assessment and conservation of medicinal plant Wealth of Haryana: 147-199. In Trivedi, P.C. (eds.): *Medicinal Plants: Ethnobotanical Approach*. Agrobios, India.

Kumbhojkar, M.S., Kulkarni, D.K., Upadhye, A.S. 1991. Ethnobotany of *Cissus quadrangularis* L. from India. *Ethnobotany* 3: 21-25.

Lalramnghinglova, H. 2003. *Ethnomedicinal Plants of Mizoram*. Bishen Singh Mahendra Pal Singh, Dehradun (India).

Lalramnghinglova, J.H. 1996. Ethnobotany of Mizoram-a preliminary survey: 439-459. In Maheshwari, J.K. (eds.): *Ethnobotany In South Asia*. Scientific Publishers, Jodhpur.

Mackay, E. 1933. Decorated carnelian beads. *Man* 33: 143-145.

Maheshwari, J.K. 1996. (ed.): *Ethnobotany in South Asia*. Scientific Publishers, Jodhpur.

Maheshwari, J.K. 2000. (ed.): *Ethnobotany and Medicinal Plants of Indian Subcontinent.* Scientific Publishers, Jodhpur.

Majumdar, D.N. 1950. *The Affairs of a Tribe*. Universal, Lucknow.

Malandhra, W.W. 1979. Atharvaveda 2,27: evidence for a soma amulet. *J. Amer. Oriental Soc.* 99: 220-222.

Maliya, S.D. 2004. Some new or less–known folk medicines of district Bahraich, Uttar Pradesh, India–II. *Ethnobotany* 16: 113-115.

Maliya, S.D. and Singh, K.K. 2003. Some new or less known folk medicines of Bahraich district, Uttar Pradesh. *Ethnobotany* 15: 132-135.

Man, E.H. 1882. On the aboriginal inhabitants of the Andamese Islands, Part III. *J. Roy. Asiat. Soc. Gr. Brit. and Ire.* 12: 329-334.

Manilal, K.S. 1989. Linkages of ethnobotany with other sciences and discipline. *Ethnobotany* 1: 15-14.

Mann, R.S. 1975. 'Change' and 'continuity' among the Nicobarese. *Eastern Anthropol.* 28: 327-339.

Manoranjan Sharma, H., Radhapyari Devi, A. and Manihar Sharma, B. 2003. Ethnomedicinal uses of monocotyledonous plants by the Meiteis of Manipur: 473-480. In Singh, V. and Jain, A.P. (eds.): *Ethnobotany and Medicinal Plants of India and Nepal* (Vol. I). Scientific Publishers, Jodhpur.

Mehra, K.L. Kanodia, K.C. and Srivastava, R.N. 1975. Folk uses of plants for Adornment in India. *Eco. Bot.* 29: 39-46.

Mills, J.P. 1926. *The Ao Nagas*. 1973 reprint. Oxford Univ. Press, Bombay.

Mishra, S. 2008. Ethnomedicinal studies of Korku tribe with Gotra names derived from plant names from Khandwa district in Madhya Pradesh. *Ethnobotany* 20: 122-127.

Mishra, S., Jha, V. and Jha, S. 1996. Plants in ethnoveterinary practices in Darbhanga (North Bihar): 189-193. In Jain, S.K. (eds.): *Ethnobiology in Human Welfare*. Deep Publications, New Delhi.

Mittre, V. 1997. Wild plants in Indian folk life-a historical perspective: 35-53. In Jain, S.K. (eds.): *Contribution to Indian Ethnobotany*. Scientific Publishers, Jodhpur.

Modi, J.J. 1893. Charms or amulets for some diseases of the eye. *J. Asiat. Soc. Bombay* 3: 338-340.

Molla, H.A. and Roy, B. 1996. Some ethnobotanical claims from the Jalpaiguri district of West Bengal: 322-324. In Maheshwari, J.K. (eds.): *Ethnobotany in South Asia.* Scientific Publishers, Jodhpur.

Mookerji, R. 1930. A study of human ornamentation. *Man in India* 10: 216-242.

Mudgal, V. 1987. Literature on ethnobotany-Indian and foreign: 45-57. In Jain, S.K. (ed.): *A Manual of Ethnobotany.* Science Publication, Jodhpur, India.

Mukerjee, R. 1944. *The Gondwana and the Gonds.* Universal, Lucknow.

Muthukumaraswamy, S., Mohan, V.R., Kumarsen, S. and Chellladurai, V. 2003. Herbal medicinal plants used by Palliyars to obtain relief from gastro-intestinal complaints: 711-714. In Singh, V. and Jain, A.P. (eds.): *Ethnobotany and Medicinal Plants of India and Nepal.* Vol II. Scientiffic Publishers, Jodhpur.

Nadkarni, K.M., 1954. *Indian Materia Medica* 3rd ed. revised and enlarged by A.K. Nadkarni Vols. 1-2. Popular Prakashan, Bombay.

Nagar, M. 1980. Pers. Comm.

Namboothiry, A.K.V. 1980. *Murugesh the Florist.* Indian Express, Bombay 26 August, p. 3.

Nargas, J. and Trivedi, P.C. 2000. Traditional and medicinal importance of *Azadirachta indica* Juss. in India: 33-37. In Maheshwari, J.K. (eds.): *Ethnobotany and Medicinal Plants of Indian Subcontinent.* Scientific Publishers, Jodhpur.

Nayar, M.P., Ramamurthy, K. and Agarwal, V.S. 1989. *Economic Plants of India.* Vols. I and II. Bot. Survey of India Calcutta.

Nazarudeen, A., Seeni, S., Koshy, K.C. and Pushpangadan, P. 1996. Folk plants of food, medicine, adornment and repellent used by the Paniyar community in North Kerala: 299-305. In Maheshwari, J.K. (eds.): *Ethnobotany in South Asia.* Scientific Publishers, Jodhpur.

Negi, C.S. 2005. Socio-cultural and ethnobotanical value of a sacred forest, *Thal Ke Dhar*, Central Himalaya. *Indian J. Trad. Knowledge* 4(2): 190-198.

Pal, D.C. 1997. Plants used in treatment of cattle and birds among tribals of Eastern India: 277-288. In Jain, S.K. (eds.): *Contribution to Indian Ethnobotany.* Scientific Publishers, Jodhpur.

Pandey, A.K., Bora, H.R. and Deka, S.C. 1996. An ethno-medico-botanical study of Golaghat district, Assam-native plant remedies for jaundice: 344-349. In Maheshwari, J.K. (eds.): *Ethnobotany in South Asia.* Scientific Publishers, Jodhpur.

Parabia, M. and Reddy, M.N. 2002. Protocol for ethnomedicinal studies: 383-395. In: Trivedi, P.C. (ed): *Ethnobotany.* Aavishkar Publications., Jaipur.

Patel, N.B., Sidana, B.S. and Jain, B.K. 2003. Tribal artifacts of Dholwani forest of Sabarkantha (Gujarat). *Ethnobotany* 15: 40-43.

Patil, S.H. and Merat, M.M. 2003. Ethnoveterinary practices in Satpudas of Nandurbar district of Maharashtra. *Ethnobotany* 15: 103-106.

Prajapati, N.S., Purohit, S.S., Sharma, A.K. and Kumar, T. 2006. *A Handbook of Medicinal Plants (A Complete Source Book).* Agrobios (India), Jodhpur.

Punjani, B.L. 2002. Plants used in contact therapy by tribals of Sabarkantha (Gujarat). *Ethnobotany* 14: 57-59.

Punjani, B.L. 2004. Ethnomedicinal uses of Vitaceae among the tribals of North Gujarat. *Ethnobotany* 16: 83-85.

Radcliffe-Brown, A.R. 1922. *The Andaman Islanders,* 1967 reprint. Free Press, New York.

Radhakrishnan, K., Pandurangan, A.G. and Pushpangadan, P. 2000. Tribal artefacts of Kerala. *Ethnobotany* 12: 67-71.

Rai, B.K., Ayachi, S.S. and Rai, A. 1996. A note on ethno-myco-medicines from Central India: 186-191. In Maheshwari, J.K. (eds.): *Ethnobotany in South Asia.* Scientific Publishers, Jodhpur.

Rajan, G.B., Khan, I.A. and Khanum, A. 2005. Role of medicinal plants in dentistry: 285-327. In Khan, I.A. and Khanum, A. (eds.): *Role of Biotechnology in Medicinal Aromatic Plants.* Vol. XII. Ukaaz Publications, Hyderabad(A.P.).

Rajendran, A. and Henry, A.N. 1994. Plants used by the tribe Kadar in Annamalai Hills of Tamil Nadu. *Ethnobotany* 6: 19-24.

Rajendran, K., Balakrishnan, R. and Chandrasekaran, S. 2003. Common medicinal plants and their utilization by villagers in East Coast districts of Tamil Nadu: 727-731. In Singh, V. and Jain, A.P. (eds.): *Ethnobotany and Medicinal Plants of India and Nepal.* Vol II. Scientific Publishers, Jodhpur.

Rajendran, S.M. and Aswal, B.S. 2000. Some flowering plants used as cosmetics among tribals of Nilgiris, Tamil Nadu, India: 425-430. In Maheshwari, J.K. (eds.): *Ethnobotany and Medicinal Plants of Indian Subcontinent.* Scientific Publishers, Jodhpur.

Raju, R.A. 2000. *Wild Plants of Indian Sub-continent and Their Economic Uses.* CBS Publishers and Distributors, New Delhi.

Rama Rao, N. and Henry, A.N. 1996. *The Ethnobotany of Eastern Ghats in Andhra Pradesh, India.* Botanical Survey of India.

Ramachandran, K., Kashyapa, K. and Chand, R. 1986. *The Useful Plants of India.* Publications and Information Directorate, CSIR. New Delhi.

Ramadas, S.R., Ghotge, N.S., Ashalata, S., Mathur, N.P., Broome,V.G. and Rao, S. 2000. Ethnoveterinary remedies used in common surgical conditions in some districts of Andhra Pradesh and Maharashtra, India. *Ethnobotany* 12: 100-112.

Ranjan, V. 1996. Some ethnomedicinal plants of Lalitpur district, Uttar Pradesh, India: 149-150. In Jain, S.K. (ed.): *Ethnobiology in Human Welfare.* Deep Publications, New Delhi.

Rao, N.S.B., Rajasekhar, D., Raju, D.C. and Nagaraju, N. 1996. Ethno-medicinal notes on some plants of Tirumala Hills for Dental disorders. *Ethnobotany* 8: 88-91.

Rao, R.R. 1997. Ethnobotanical studies on some adivasi tribes of North East India with special reference to the Naga people: 209-223. In Jain, S.K. (eds.): *Contribution to Indian Ethnobotany*. Scientific Publishers, Jodhpur.

Rastogi, R.P. and Mehrotra, B.N. 1990. *Compendium of Indian Medicinal Plants*. Vol. 1. (1960-1969). CDRI and CSIR.

Rastogi, R.P. and Mehrotra, B.N. 1991. *Compendium of Indian Medicinal Plants*. Vol. 2. (1970-1979). CDRI and CSIR.

Rastogi, R.P. and Mehrotra, B.N. 1993. *Compendium of Indian Medicinal Plants*. Vol. 3. (1980-1984). CDRI and CSIR.

Rastogi, R.P. and Mehrotra, B.N. 1995a. *Compendium of Indian Medicinal Plants*. Vol. 4. (1985-1989). CDRI and CSIR.

Rastogi, R.P. and Mehrotra, B.N. 1995b. *Compendium of Indian Medicinal Plants*. Vol. 5. (1990-1994). CDRI and CSIR.

Ravikumar, K. and Vijay Sankar, R. 2003. Ethnobotany of Malayali tribes in Melpattu village, Javvadhu Hills of Eastern Ghats, Tiruvannamalai District, Tamil Nadu: 715-726. In Singh, V. and Jain, A.P. (eds.): *Ethnobotany and Medicinal Plants of India and Nepal*. Vol II. Scientific Publishers, Jodhpur.

Rawat, M.S. and Chowdhury, S. 1998. *Ethno-Medico-Botany of Arunachal Pradesh*. Bishen Singh Mahendra Pal Singh, Dehradun (India).

Reddy, N.S. 1956. Rites and customs associated with marriage in a north Indian village, Part 1 and 2. *Eastern Anthropol.* 9: 77-86; 179-184.

Roy, S.C. 1925. *The Bihors*. Mission Press, Ranchi.

Sahoo, A.K. and Bahali, D.D. 2003. Plants used in less known ethnomedicines and medico-religious beliefs in Phulbani district of Orissa: 500-504. In Singh, V. and Jain, A.P. (eds.): *Ethnobotany and Medicinal Plants of India and Nepal*. Vol. I. Scientific Publishers, Jodhpur.

Saini, D.C. 1996. Ethnobotany of Tharus of Basti District, Uttar Pradesh: 138-153. In Maheshwari, J.K. (eds.): *Ethnobotany in South Asia*. Scientific Publishers, Jodhpur.

Saini, V.K. 1996. Plants in the welfare of tribal women and children in certain areas of Central India: 140-144. In Jain, S.K. (eds.): *Ethnobiology in Human Welfare*. Deep Publications, New Delhi.

Saklani, A. and Jain, S.K. 1994. *Cross-Cultural Ethnobotany of North East India*. Deep Publications, New Delhi.

Sarangi, N. and Sahu, R.K. 2004. Ethno-medicinal plants used in venereal and gynaecological disorders in Kalahandi, Orissa. *Ethnobotany* 16: 16-20.

Saren, A.M., Sen, R. and Pal, D.C. 2000. A contribution to the ethnobotany of Bankura district, West Bengal: 545-555. In Maheshwari, J.K. (eds.): *Ethnobotany and Medicinal Plants of Indian Subcontinent*. Scientific Publishers, Jodhpur.

Sastry, B.L.N. 1966. Cosmetics through the early ages with special reference to Buddhist period. *J. Andhra Hist. Res. Soc.* 31: 45-48.

Saxena, H.O., Brahmam, M. and Dutta, P.K. 1997. Ethnobotanical studies in Orissa: 123-1135. In Jain, S.K. (eds.): *Contribution to Indian Ethnobotany*. Scientific Publishers, Jodhpur.

Schoff, W.H. 1922. Aloes. *J. Amer. Oriental. Soc.* 42: 172-185.

Schultes, R.E. 1962. The role of ethnobotanists in the search for new medicinal plants. *Lloydia* 35: 257-266.

Sebestian, M.K. and Bhandari, M.M. 1984. Magico-religious beliefs about plants among the Bhils of Udaipur district of Rajasthan. *Folklore* April: 77-80.

Seetharami Reddi, T.V.V., Prasanthi, S. and Ramarao Naidu, B.V.A. 2005. Medicinal and aromatic plants of India: 1-146. In Khan, I.A. and Khanum, A. (eds.): *Role of Biotechnology in Medicinal and Aromatic Plants.* Vol. XII. Ukaaz Publications, Hyderabad (A.P.).

Sen, P., Dollo, M., Choudhury, M.D. and Choudhury, D. 2008. Documentation of traditional herbal knowledge of Khamptis of Arunachal Pradesh. *Indian J. Trad. Knowledge* 7 (3): 438-442.

Sen, S. and Batra, A. 1997. Ethno-medico-botany of household remedies of Phagi tehsil of Jaipur district (Rajasthan). *Ethnobotany* 9: 122-128.

Sen, S.K. and Behera, L.M. 2007. Ethnomedicinal plants used in touch therapy at Bargarh district of Orissa. *Ethnobotany* 19: 100-104.

Sen Gupta, S. 1980. *Sacred Trees Across Cultures and Nations.* Indian Publications, Kolkata.

Shah, N.C. and Singh, S.C. 1990. Hitherto unreported phytotherapeutical uses from tribal pockets of Madhya Pradesh (India). *Ethnobotany* 2: 91-95.

Sharma, B.D. and Rana, J.C. 2005. *Plant Genetic Resources of Western Himalayas.* Bishen Singh Mahendra Pal Singh, Dehradun (India).

Sharma, M.H., Devi, A.R. and Sharma, B.M. 2003. Ethnomedicinal uses of monocotyledonous plants by the Meiteis of Manipur: 473-480. In Singh, V. and Jain, A.P. (eds.): *Ethnobotany and Medicinal Plants of India and Nepal.* Vol. I. Scientific Publishers, Jodhpur.

Sharma, N. 2003. Medicinal uses of macrofungi. *Ethnobotany* 15: 97-99.

Sharma, N.K. 2002. Ethno-medico-religious plants of Hadoti Plateau. (S.E. Rajasthan)- a preliminary survey: 394-411. In Trivedi, P.C. (eds.): *Ethnobotany*. Aavishkar Publishers, Distributors, Jaipur.

Sharma, P.P. and Singh, N.P. 2000. Less known ethnobotanical uses of plants in Dadra and Nagar Haveli (U.T.): Part II (L-Z). *Ethnobotany* 12: 81-85.

Sharma, R. 2003. *Medicinal Plants of India, An Encyclopedia*. Daya Publishing House, Delhi.

Sharma, S.K. 2000. Plants used as Henna dye by Bhils of Southern Rajasthan: 257. In Maheshwari, J.K. (eds.): *Ethnobotany and Medicinal Plants of Indian Subcontinent*. Scientific Publishers, Jodhpur.

Shende, N.J. 1949. The foundations of the Atharvanic religion. *Bull, Deccan College Res, Inst*. 9: 197-412.

Sikarwar, R.L.S. 1996. Ethno-veterinary herbal medicines in Morena district of Madhya Pradesh, India: 194-196. In Jain, S.K. (eds.): *Ethnobiology in Human Welfare*. Deep Publications, New Delhi.

Sikarwar, R.L.S. 2002. Ethnogynaecological uses of plants new to India. *Ethnobotany* 14: 112-115.

Sikarwar, R.L.S. and Kaushik, J.P. 1992. Some less known medicinal uses of trees among the Sahariyas of Morena district, M.P., India. *Ethnobotany* 4: 71-74.

Singh, A.K. 2000. A contribution to the ethnobotany of sub-Himalayan region of Eastern Uttar Pradesh: 237-246. In Maheshwari, J.K. (eds.): *Ethnobotany and Medicinal Plants of Indian Subcontinent*. Scientific Publishers, Jodhpur.

Singh, K.K. and Kumar, K. 2000. *Ethnobotanical Wisdom of Gaddi Tribe in Western Himalaya*. Bishen Singh Mahendra Pal Singh, Dehradun (India).

Singh, K.K. and Maheshwari, J.K. 1992. Folk medicinal uses of some plants among the Tharus of Gorakhpur district, Uttar Pradesh, India. *Ethnobotany* 4: 39-43.

Singh, K.K. and Prakash, A. 1994. Indigenous phytotherapy among the Gond tribe of Uttar Pradesh, India. *Ethnobotany* 6: 37-41.

Singh, S.K. and Rawat, G.S. 2000. *Flora of Great Himalayan National Park, Himachal Pradesh*. Bishen Singh Mahendra Pal Singh, Dehradun (India).

Singh, U., 1996. *Dictionary of Economic Plants in India*. I.C.A.R., New Delhi.

Singh, U. and Narain, S. 2008. A short note on traditional remedy through plant wreath. *Ethnobotany* 20: 152-153.

Singh, V. and Pandey, R.P. 1998. *Ethnobotany of Rajasthan, India*. Scientific Publishers, (India).

Sinha, R.K and Sinha, S. 2001. *Ethnobiology*. Surabhi Publications, Jaipur.

Siwakoti, M. and Varma, S.K. 1996. Medicinal plants of the Terai of Eastern Nepal: 423-438. In Maheshwari, J.K. (eds.): *Ethnobotany in South Asia*. Scientific Publishers, Jodhpur.

Sood, S.K., Nath, R. and Kalia, D.C. 2001. *Ethnobotany of Cold Desert Tribes of Lahoul-Spiti (N.W. Himalaya)*. Deep Publications, New Delhi.

Sood, S.K., Negi, C.J. and Lakhanpal, T.N. 2006. *Orchidaceae and Mankind*. Deep Publication, New Delhi.

Sood, S.K., Rawat, S. and Rawat, D. 2009. *Dye, Masticatory and Fumitory Resources of India*. International Book Distributors, Dehradun, Uttarakhand (India).

Sood, S.K., Sharma, S.K., Kumar, N. and Kumar, H. 2009. *Ethnobotanical studies on Trees, Shrubs and Climbers of Himalaya.* Satish Serial Publishing House, Azadpur, Delhi.

Sood, S.K. and Sudershana. 2008. *Root Drug Plants of Ethnic India.* Satish Serial Publishing House, Azadpur, Delhi.

Sood, S.K. and Thakur, S. 2004. *Ethnobotany of Rewalsar Himalaya.* Deep Publications, New Delhi.

Sood, S.K., Thakur, V. and Lakhanpal, T.N. 2005. *Sacred and Magico-Religious Plants of India.* Scientific Publishers, India (Jodhpur).

Subramaniam, A. 2000. A survey of medicinal plants from Chitheri Hills in Dharmapuri district, Tamil Nadu: 395-416. In Maheshwari, J.K. (eds.): *Ethnobotany and Medicinal Plants of Indian Subcontinent.* Scientific Publishers, Jodhpur.

Sundararaj, D. 1959. *Guide to Economic Plants of South India.* Periodical Expert Book Agency, Delhi.

Tarafder, C.R., and Rai, Chaudhuri, H.N. 1997. Less known medicinal uses of plants among the tribals of Hazaribagh district of Bihar: 93-101. In Jain, S.K. (eds.): *Contribution to Indian Ethnobotany.* Scientific Publishers, Jodhpur.

The Wealth of India. 1948-1973. *Raw Materials,* Vol. I-X. CSIR, New Delhi.

Thomas, S., Haridarshan, K. and Borthakur, S.K. 1998. Ethnobotanical observations on Rattan Palms among Adi and Nishing tribes in Arunachal Pradesh. *Ethnobotany* 10: 22-26.

Tosh, J. 1996. Ethnobotanical study of Western Maharashtra: 169-174. In Maheshwari, J.K. (eds.): *Ethnobotany in South Asia.* Scientific Publishers, Jodhpur.

Tripathi, S. and Goel, A.K. 2001. Ethnobotanical diversity of Zingiberaceae in North-Eastern India. *Ethnobotany* 13: 67-79.

Uniyal, M.R. 2007. Medicinal plants of India: need for their preservation: 11-15. In Chopra, A.K., Khanna, D.R., Prasad, G., Malik, D.S. and Bhutiani, R. (eds.): *Medicinal Plants: Conservation, Cultivation, and Utilization.* Daya Publishers, Delhi.

Upadhaya, A.S., Kumbhojkar, M.S. and Kulkarni, D.K. 1997. Ethnomedicobotany of some sacred plants of Western Maharashtra. *Ethnobotany* 9: 65-66.

Upadhye, A.S. and Kumbhojkar, M.S. 2003. Ethnobotany of *Butea monosperma*(Lam.) Taub: 94-98. In Singh, V. and Jain, A.P. (eds.): *Ethnobotany and Medicinal Plants of India and Nepal.* (Vol I). Scientific Publishers, Jodhpur.

Varghese, E. 1996. *Applied Ethnobotany.* Deep Publications, New Delhi.

Vidyarthi, L.P. 1975. The Future of Traditional 'primitive' societies: a case study of an Indian shifting cultivation society. *Eastern Anthropol.* 28: 313-326.

Vihari, V. 1995. Ethnobotany of cosmetics of Indo-Nepal border. *Ethnobotany* 7: 89-94.

Wasson, R.G. 1971. The soma of the Rig Veda: what was it? *J. Amer. Oriental Soc.* 91: 169-186.

Watt, G. 1889-1896. *A Dictionary of the Economic Products of India*. Vol. I-VI. Periodical Experts, New Delhi.

Watt, G. 1972. *Dictionary of the Economic Products of India*. Jayed Press, Ballimaran, Delhi.

Appendices

Appendix–I
Index to Plants Used for Adornments, Touch Therapy and Tattooing Under Various Divisions of Plant Kingdom

Division	*No. of Families*	*Total No. of Genera/ Species*	*Total No. of Species in a Genus*
Alga	1	1/1	*Caulerpa racemosa* (1).
Fungi	3	3/3	*Agaricus* (1), *Amanita* (1), *Tuber* (1).
Gymnosperms	3	3/4	*Gnetum* (1), *Pinus* (2), *Taxus* (1).
Pteridophytes	9	9/11	*Actinopteris* (1), *Adiantum* (1), *Asplenium* (1), *Equisetum* (1), *Helminthostachys* (1), *Lygodium* (2), *Marsilea* (1), *Selaginella* (2), *Thelypteris* (1).
Dicots	93	275/343	*Abelmoschus* (1), *Abroma* (1), *Abrus* (1), *Abutilon* (1), *Acacia* (5), *Achillea* (1), *Achyranthes* (1), *Adenanthera* (1), *Aegiceras* (1), *Aegle* (1), *Aeschynomene* (2), *Ageratum* (1), *Ajuga* (1), *Albizia* (1), *Aleurites* (1), *Alstonia* (1), *Alternanthera* (1), *Alysicarpus* (1), *Ammania* (2), *Ampelocissus* (1), *Anisophyllea* (1), *Annona* (1), *Anogeissus* (1), *Anthocephalus* (1), *Antidesma* (1), *Aporusa* (1), *Aquilaria* (1), *Argemone* (1), *Argyreia* (1), *Arnebia* (2), *Artemisia* (1), *Atylosia* (1), *Azadirachta* (1), *Baccaurea* (2), *Baphicacanthus* (1), *Barleria* (1), *Basella* (1), *Bauhinia* (2), *Begonia* (1), *Berberis* (1), *Bergenia* (1), *Biophytum* (2), *Bixa* (1), *Blastania* (1), *Boerhaavia* (1), *Bombax* (1), *Borreria* (1), *Boswellia* (1), *Bougainvillaea* (1), *Brassica* (1),*Bryophyllum* (1), *Butea* (1), *Caesalpinia* (1), *Cajanus* (1), *Callicarpa* (2), *Calophyllum* (1), *Calotropis* (2), *Cannabis* (1), *Capparis* (2), *Capsicum* (1), *Cardiospermum* (1), *Careya* (1), *Carica* (1), *Carissa* (1), *Cascabela* (1), *Casearia* (2), *Cassia* (3), *Cassytha* (1), *Catharanthus* (1), *Cayratia* (1), *Celastrus* (1), *Celosia* (1), *Celtis* (1), *Ceriscoides* (1), *Chenopodium* (1), *Cirsium* (1), *Cissampelos* (1), *Cissus* (1), *Citrus* (3), *Claoxylon* (1), *Cleome* (1), *Clerodendrum* (4), *Clitoria* (1), *Cochlospermum* (1), *Codiaeum* (1), *Congea* (1), *Corchorus* (1), *Cordia* (1), *Crataeva* (1), *Crossandra* (1), *Crotalaria* (1), *Croton* (2), *Dalbergia* (1), *Daphne* (1), *Datura* (2),

Contd..

Appendix–I–*Contd...*

Division	*No. of Families*	*Total No. of Genera/ Species*	*Total No. of Species in a Genus*
			Delonix (1), *Dendrophthoe* (1), *Diplocyclos* (1), *Drypetes* (1), *Echinops* (1), *Eclipta* (1), *Elaeagnus* (1), *Elaeocarpus* (4), *Elephantopus* (1), *Elytraria* (1), *Emilia* (1), *Enicostema* (1), *Entada* (1), *Ervatamia* (1), *Erythrina* (2), *Eucalyptus* (1), *Euclea* (1), *Eugenia*(1), *Euonymus*(2), *Euphorbia* (2), *Evolvulus* (1), *Ficus* (4), *Flacourtia* (1), *Flemingia* (2), *Flueggea* (1), *Garcinia* (1), *Gardenia* (1), *Geranium* (1), *Gnidia* (1), *Gossypium* (1), *Grewia* (1), *Guazuma* (1), *Guettarda* (1), *Gyrocarpus* (1), *Haldina* (1), *Helianthus* (1), *Helicteres* (1), *Hemidesmus* (1), *Hevea* (1), *Hibiscus* (2), *Hiptage* (1), *Holarrhena* (1), *Holoptelea* (1), *Hydrocotyle* (1), *Ichnocarpus* (1), *Impatiens* (2), *Indigofera* (3), *Ipomoea* (2), *Ixora* (2), *Jacaranda* (1), *Jasminum* (7), *Jatropha* (1), *Juglans* (1), *Justicia* (3), *Lablab* (1), *Lagenaria* (1), *Lannea* (1), *Launaea* (1), *Lawsonia* (1), *Leea* (1), *Leucaena* (1), *Leucas* (2), *Linum* (1), *Litsea* (1), *Lobelia* (1), *Luffa* (1), *Madhuca* (1), *Magnolia* (1), *Mallotus* (2), *Mangifera* (1), *Manilkara* (1), *Melastoma* (1), *Melia* (1), *Melothria* (2), *Mesua* (1), *Michelia* (1), *Millettia* (1), *Mimosa* (1), *Mimusops* (1), *Mirabilis* (1), *Momordica* (1), *Mucuna* (1), *Murraya* (2), *Mussaenda* (1), *Myristica*(1), *Naravelia* (1), *Naringi* (1), *Nelumbo* (1), *Neolitsea* (1), *Nerium* (1), *Nicotiana* (1), *Nyctanthes* (1), *Nymphaea* (1), *Ocimum* (2), *Opilia* (1), *Oroxylum* (1), *Osbeckia* (1), *Oxalis* (1), *Paedaria* (1), *Papaver* (1), *Passiflora* (1), *Pavetta* (1), *Phyla* (1), *Phyllanthus* (1), *Phyllocephalum* (1), *Piper* (1), *Pithecellobium* (1), *Plumbago* (2), *Pogostemon* (1), *Pongamia* (1), *Porana* (1), *Potentilla* (1), *Premna* (1), *Prunus* (2), *Psychotria* (1), *Pterocarpus* (2), *Punica* (1), *Quisqualis* (1), *Ricinus* (1), *Rosa* (2), *Rubia* (1), *Rubus* (2), *Rumex* (1), *Samadera* (1), *Sambucus* (1), *Santalum* (1), *Sapindus* (1), *Saraca* (1), *Schima* (1), *Schrebera* (1), *Scoparia* (1), *Scurrula* (1), *Securinega* (1), *Semecarpus* (1), *Shorea* (1), *Shuteria* (1), *Sida* (1), *Smithia* (1), *Solanum* (3), *Spermadictyon* (1), *Sphaeranthus* (1), *Spondias* (1),

Contd..

Appendix–I–*Contd...*

Division	*No. of Families*	*Total No. of Genera/ Species*	*Total No. of Species in a Genus*
			Stephania (2), *Stereospermum* (1), *Streblus* (1), *Striga* (1), *Symphorema* (1), *Symplocos* (2), *Syzygium* (1), *Tabernaemontana* (1), *Tadehagi* (1), *Tagetes* (2), *Tamarindus* (1), *Tamarix* (2), *Taraxacum* (1), *Tectona* (1), *Tephrosia* (1), *Terminalia* (4), *Thunbergia* (1), *Tinospora* (1), *Toddalia* (1), *Tragopogon* (1), *Tribulus* (1), *Trichodesma* (1), *Tridax* (1), *Trigonella* (1), *Tylophora* (1), *Uraria* (1), *Urena* (1), *Urtica* (1), *Vateria* (1), *Vernonia* (1), *Vitex* (1), *Wattakaka* (1), *Woodfordia* (1), *Wrightia* (2), *Xanthium* (1), *Xeromphis* (1), *Zanthoxylum* (1), *Zizyphus* (2).
Monocots	12	56/78	*Acorus* (1), *Agave* (1), *Allium* (2), *Aloe* (1), *Alpinia* (1), *Anadendrum* (1), *Areca* (1), *Asparagus* (2), *Bambusa* (1), *Borassus* (2), *Bulbophyllum* (1), *Caladium* (1), *Calamus* (4), *Canna* (2), *Caryota* (2), *Chlorophytum* (1), *Cocos* (1), *Coelogyne* (1), *Coix* (3), *Commelina* (1), *Corypha* (1), *Crinum* (2), *Curculigo* (1), *Curcuma* (3), *Cynodon* (1), *Dendrobium* (2), *Dendrocalamus* (1), *Desmostachys* (1), *Dichrostachys* (1), *Didymosperma* (1), *Elettaria* (1), *Ensete* (1), *Globba* (1), *Gloriosa* (1), *Hordeum* (1), *Hornstedtia* (1), *Hymenachne* (1), *Kaempferia* (1), *Licuala* (1), *Livistonia* (2), *Molineria* (1), *Musa* (2), *Nannorhops* (1), *Nypa* (1), *Oryza* (1), *Pandanus* (2), *Phoenix* (2), *Phragmites* (1), *Pothos* (1), *Rhynchostylis* (1), *Saccharum* (3), *Smilax* (2), *Thysanolaena* (1), *Triticum* (1), *Vanda* (2), *Vetiveria* (1).

Appendix-II
Index to Families of Adornments, Touch Therapy and Tattooing

Name of Family	*Genus/Species*	*Total No. of Species under a Genus*
Acanthaceae	6/8	*Baphicacanthus* (1), *Barleria* (1), *Crossandra* (1), *Elytraria* (1), *Justicia* (3), *Thunbergia* (1)
Actinopteridaceae	1/1	*Actinopteris* (1)
Adiantaceae	1/1	*Adiantum* (1)
Agaricaceae	1/1	*Agaricus* (1)
Agavaceae	1/1	*Agave* (1)
Amanitaceae	1/1	*Amanita* (1)
Amaranthaceae	3/3	*Achyranthes* (1), *Alternanthera* (1), *Celosia* (1)
Amaryllidaceae	2/3	*Crinum* (2), *Curculigo* (1)
Anacardiaceae	4/4	*Lannea* (1), *Mangifera* (1), *Semecarpus* (1), *Spondias* (1)
Annonaceae	1/1	*Annona* (1)
Apiaceae	1/1	*Hydrocotyle* (1)
Apocynaceae	10/11	*Alstonia* (1), *Carissa* (1), *Cascabela* (1), *Catharanthus* (1), *Ervatamia* (1), *Holarrhena* (1), *Ichnocarpus* (1), *Nerium* (1), *Tabernaemontana* (1), *Wrightia* (2).
Araceae	5/5	*Acorus* (1), *Anadendrum* (1), *Caladium* (1), *Dichrostachys* (1), *Pothos* (1)
Arecaceae	12/19	*Areca* (1), *Borassus* (2), *Calamus* (4), *Caryota* (2), *Cocos* (1), *Corypha* (1), *Didymosperma* (1), *Licula* (1), *Livistona* (2), *Nannorhops* (1), *Nypha* (1), *Phoenix* (2)
Asclepiadaceae	4/5	*Calotropis* (2), *Hemidesmus* (1), *Tylophora* (1), *Wattakaka* (1)
Aspleniaceae	1/1	*Asplenium* (1)
Asteraceae	18/19	*Achillea* (1), *Ageratum* (1), *Artemisia* (1), *Cirsium* (1), *Echinops* (1), *Eclipta* (1), *Elephantopus* (1), *Emilia* (1), *Helianthus* (1), *Launaea* (1), *Phyllocephalum* (1), *Sphaeranthus* (1), *Tagetes* (2), *Taraxacum* (1), *Tragopogon* (1), *Tridax* (1), *Vernonia* (1), *Xanthium* (1)
Balsaminaceae	1/2	*Impatiens* (2)

Contd...

Appendix–II–*Contd...*

Name of Family	*Genus/Species*	*Total No. of Species under a Genus*
Begoniaceae	1/1	*Begonia* (1)
Berberidaceae	1/1	*Berberis* (1)
Bignoniaceae	3/3	*Jacaranda* (1), *Oroxylum* (1), *Stereospermum* (1).
Bixaceae	1/1	*Bixa* (1)
Bombaceae	1/1	*Bombax* (1)
Boraginaceae	2/3	*Arnebia* (2), *Trichodesma* (1)
Brassicaceae	1/1	*Brassica* (1)
Burseraceae	1/1	*Boswellia* (1)
Campanulaceae	1/1	*Lobelia* (1)
Cannabinaceae	1/1	*Cannabis* (1)
Cannaceae	1/2	*Canna* (2)
Capparidaceae	3/4	*Capparis* (2), *Crataeva* (1), *Cleome* (1)
Caprifoliaceae	1/1	*Sambucus* (1)
Caricaceae	1/1	*Carica* (1)
Celastraceae	2/3	*Celastrus* (1), *Euonymus* (2)
Chenopodiaceae	2/2	*Basella* (1), *Chenopodium* (1)
Chlorophyceae	1/1	*Caulerpa* (1)
Clusiaceae	3/3	*Calophyllum* (1), *Garcinia* (1), *Mesua* (1)
Cochlospermaceae	1/1	*Cochlospermum* (1)
Combretaceae	4/7	*Anogeissus* (1), *Gyrocarpus* (1), *Quisqualis* (1), *Terminalia* (4)
Commelinaceae	1/1	*Commelina* (1)
Convolvulaceae	4/5	*Argyreia* (1), *Evolvulus* (1), *Ipomoea* (2), *Porana* (1).
Crassulaceae	1/1	*Bryophyllum* (1)
Cucurbitaceae	6/7	*Blastania* (1), *Diplocyclos* (1), *Lagenaria* (1), *Luffa* (1), *Melothria* (2), *Momordica* (1)
Dipterocarpaceae	2/2	*Shorea* (1), *Vateria* (1)
Ebenaceae	1/1	*Euclea* (1)
Ehretiaceae	1/1	*Cordia* (1)
Elaeagnaceae	1/1	*Elaeagnus* (1)
Elaeocarpaceae	1/4	*Elaeocarpus* (4)

Contd...

Appendix–II–*Contd...*

Name of Family	*Genus/Species*	*Total No. of Species under a Genus*
Equisetaceae	1/1	*Equisetum* (1)
Euphorbiaceae	16/20	*Aleurites* (1), *Antidesma* (1), *Aporusa* (1), *Baccaurea* (2), *Claoxylon* (1), *Codiaeum* (1), *Croton* (2), *Drypetes* (1), *Euphorbia* (2), *Flueggea* (1), Hevea (1), *Jatropha* (1), *Mallotus* (2), *Phyllanthus* (1), *Ricinus* (1), *Securinega* (1)
Fabaceae	36/49	*Abrus* (1), *Acacia* (5), *Adenanthera* (1), *Aeschynomene* (2), *Albizia* (1), *Alysicarpus* (1), *Atylosia* (1), *Bauhinia* (2), *Butea* (1), *Caesalpinia* (1), *Cajanus* (1), *Cassia* (3), *Clitoria* (1), *Crotalaria* (1), *Dalbergia* (1), *Delonix* (1), *Entada* (1), *Erythrina* (2), *Flemingia* (2), *Indigofera* (3), *Lablab* (1), *Leucaena* (1), *Millettia* (1), *Mimosa* (1), *Mucuna* (1), *Pithecellobium* (1), *Pongamia* (1), *Pterocarpus* (2), *Saraca* (1), *Shuteria* (1), *Smithia* (1), *Tadehagi* (1), *Tamarindus* (1), *Tephrosia* (1), *Trigonella* (1), *Uraria* (1)
Flacourtiaceae	2/3	*Casearia* (2), *Flacourtia* (1)
Gentianaceae	1/1	*Enicostemma* (1)
Geraniaceae	1/1	*Geranium* (1)
Gnetaceae	1/1	*Gnetum* (1)
Juglandaceae	1/1	*Juglans* (1)
Lamiaceae	4/6	*Ajuga* (1), *Leucas* (2), *Ocimum* (2), *Pogostemon* (1).
Lauraceae	3/3	*Cassytha* (1), *Litsea* (1), *Neolitsea* (1)
Leeaceae	1/1	*Leea* (1)
Liliaceae	7/10	*Allium* (2), *Aloe* (1), *Asparagus* (2), *Chlorophytum* (1), *Gloriosa* (1), *Molineria* (1), *Smilax* (2)
Linaceae	1/1	*Linum* (1)
Loranthaceae	2/2	*Dendrophthoe* (1), *Scurrula* (1)
Lythraceae	3/4	*Ammania* (2), *Lawsonia* (1), *Woodfordia* (1)
Magnoliaceae	2/2	*Magnolia* (1), *Michelia* (1)
Malpighiaceae	1/1	*Hiptage* (1)
Malvaceae	6/7	*Abelmoschus* (1), *Abutilon* (1), *Gossypium* (1), *Hibiscus* (2), *Sida* (1), *Urena* (1)
Marsileaceae	1/1	*Marsilea* (1)
Melastomataceae	2/2	*Melastoma* (1), *Osbeckia* (1)

Contd...

Appendix–II–*Contd...*

Name of Family	*Genus/Species*	*Total No. of Species under a Genus*
Meliaceae	2/2	*Azadirachta* (1), *Melia* (1)
Menispermaceae	3/4	*Cissampelos* (1), *Stephania* (2), *Tinospora* (1)
Moraceae	2/5	*Ficus* (4), *Streblus* (1)
Musaceae	2/3	*Ensete* (1), *Musa* (2)
Myristicaceae	1/1	*Myristica* (1)
Myrsinaceae	1/1	*Aegiceras* (1)
Myrtaceae	3/3	*Eucalyptus* (1), *Eugenia* (1), *Syzygium* (1)
Nyctaginaceae	3/3	*Boerhaavia* (1), *Bougainvillaea* (1), *Mirabilis* (1).
Nymphaeaceae	2/2	*Nelumbo* (1), *Nymphaea* (1)
Oleaceae	3/9	*Jasminum* (7), *Nyctanthes* (1), *Schrebera* (1)
Ophioglossaceae	1/1	*Helminthostachys* (1)
Opiliaceae	1/1	*Opilia* (1)
Orchidaceae	5/7	*Bulbophyllum* (1), *Coelogyne* (1), *Dendrobium* (2), *Rhynchostylis* (1), *Vanda* (2)
Oxalidaceae	2/3	*Biophytum* (2), *Oxalis* (1)
Pandanaceae	1/2	*Pandanus* (2)
Papaveraceae	2/2	*Argemone* (1), *Papaver* (1)
Passifloraceae	1/1	*Passiflora* (1)
Pinaceae	1/2	*Pinus* (2)
Piperaceae	1/1	*Piper* (1)
Plumbaginaceae	1/2	*Plumbago* (2)
Poaceae	13/17	*Bambusa* (1), *Coix* (3), *Cynodon* (1), *Dendrocalamus* (1), *Desmostachya* (1), *Hordeum* (1), *Hymenachne* (1), *Oryza* (1), *Phragmites* (1), *Saccharum* (3), *Thysanolaena* (1), *Triticum* (1), *Vetiveria* (1)
Polygonaceae	1/1	*Rumex* (1)
Punicaceae	1/1	*Punica* (1)
Ranunculaceae	1/1	*Naravelia* (1)
Rhamnaceae	1/2	*Zizyphus* (2)
Rhizophoraceae	1/1	*Anisophyllea* (1)
Rosaceae	4/7	*Potentilla* (1), *Prunus* (2), *Rosa* (2), *Rubus* (2)

Contd...

Appendix–II–*Contd...*

Name of Family	*Genus/Species*	*Total No. of Species under a Genus*
Rubiaceae	14/15	*Anthocephalus* (1), *Borreria* (1), *Ceriscoides* (1), *Gardenia* (1), *Guettarda* (1), *Haldina* (1), *Ixora* (2), *Mussaenda* (1), *Paederia* (1), *Pavetta* (1), *Psychotria* (1), *Rubia* (1), *Spermadictyon* (1), *Xeromphis* (1).
Rutaceae	6/9	*Aegle* (1), *Citrus* (3), *Murraya* (2), *Naringi* (1), *Toddalia* (1), *Zanthoxylum* (1)
Santalaceae	1/1	*Santalum* (1)
Sapindaceae	2/2	*Sapindus* (1), *Cardiospermum* (1)
Sapotaceae	3/3	*Madhuca* (1), *Manilkara* (1), *Mimusops* (1)
Saxifragaceae	1/1	*Bergenia* (1)
Schizaeaceae	1/2	*Lygodium* (2)
Scrophulariaceae	2/2	*Scoparia* (1), *Striga* (1)
Selaginellaceae	1/2	*Selaginella* (2)
Simarubaceae	1/1	*Samadera* (1)
Solanaceae	4/7	*Capsicum* (1), *Datura* (2), *Nicotiana* (1), *Solanum* (3).
Sterculiaceae	3/3	*Abroma* (1), *Guazuma* (1), *Helicteres* (1)
Symplocaceae	1/2	*Symplocos* (2)
Tamaricaceae	1/2	*Tamarix* (2)
Taxaceae	1/1	*Taxus* (1)
Theaceae	1/1	*Schima* (1)
Thelypteridaceae	1/1	*Thelypteris* (1)
Thymelaeaceae	3/3	*Aquilaria* (1), *Daphne* (1), *Gnidia* (1)
Tiliaceae	2/2	*Corchorus* (1), *Grewia* (1)
Tuberaceae	1/1	*Tuber* (1)
Ulmaceae	2/2	*Celtis* (1), *Holoptelea* (1)
Urticaceae	1/1	*Urtica* (1)
Verbenaceae	8/12	*Callicarpa* (2), *Clerodendrum* (4), *Congea* (1), *Phyla* (1), *Premna* (1), *Symphorema* (1), *Tectona* (1), *Vitex* (1)
Vitaceae	3/3	*Ampelocissus* (1), *Cayratia* (1), *Cissus* (1)
Zingiberaceae	6/8	*Alpinia* (1), *Curcuma* (3), *Elettaria* (1), *Globba* (1), *Hornstedtia* (1), *Kaempferia* (1)
Zygophyllaceae	1/1	*Tribulus* (1)

Appendix-III
Plants Used for Adornments

Plant Species	*Page No.*	*Plant Species*	*Page No.*
Abrus precatorius	8	*Asplenium aethiopicum*	31
Abutilon indicum	8	*Azadirachta indica*	32
Acacia catechu	9	*Baccaurea courtallensis*	32
Acacia ferruginea	9	*Bambusa vulgaris*	33
Acacia pennata	11	*Barleria prionitis*	34
Achillea millefolium	11	*Basella alba*	34
Achyranthes aspera	12	*Bauhinia racemosa*	35
Acorus calamus	13	*Bauhinia vahlii*	35
Actinopteris radiata	13	*Begonia picta*	35
Adenanthera pavonina	14	*Bixa orellana*	37
Aegiceras corniculatum	15	*Blastania garcinii*	38
Aegle marmelos	15	*Borassus flabellifer*	39
Aeschynomene aspera	16	*Borassus ramiflora*	39
Aeschynomene indica	16	*Bougainvillaea spectabilsis*	40
Agave americana	17	*Bulbophyllum neilgherrense*	41
Albizia lebbeck	18	*Butea monosperma*	42
Aleurites moluccana	18	*Caesalpinia bonduc*	42
Alpinia allughas	21	*Cajanus cajan*	43
Alstonia scholaris	21	*Caladium bicolor*	43
Amanita muscaria	22	*Calamus erectus*	43
Anadendrum paniculatum	23	*Calamus floribundus*	44
Anisophyllea zeylanica	24	*Calamus inermis*	44
Anogeissus latifolia	24	*Calamus latifolia*	44
Anthocephalus chinensis	25	*Callicarpa arborea*	44
Aquilaria agallocha	26	*Callicarpa macrophylla*	45
Areca catechu	27	*Calophyllum inophyllum*	45
Arnebia benthamii	28	*Calotropis gigantea*	46
Arnebia guttata	29	*Canna orientalis*	47
Asparagus racemosus	30	*Canna speciosa*	48

Contd..

Appendix–III–*Contd...*

Plant Species	*Page No.*	*Plant Species*	*Page No.*
Cannabis sativa	48	*Cordia gharaf*	69
Capparis aphylla	48	*Corypha umbraculifera*	70
Capsicum annuum	49	*Crataeva magna*	70
Carissa spinarum	51	*Crinum asiaticum*	71
Caryota mitis	52	*Crinum latifolium*	71
Caryota urens	52	*Crossandra undulaefolia*	72
Cascabela thevetia	52	*Crotalaria spectabilis*	72
Casearia graveolens	53	*Croton caudatus*	72
Catharanthus roseus	56	*Croton roxburghii*	73
Caulerpa racemosa	57	*Curcuma neilgherrensis*	74
Celastrus paniculatus	57	*Curcuma zedoaria*	74
Celosia argentea	58	*Cynodon dactylon*	75
Celtis cinnamomea	58	*Dalbergia sissoo*	75
Ceriscoides turgida	58	*Daphne papyracea*	76
Cissus quadrangularis	61	*Delonix regia*	77
Citrus maxima	61	*Dendrobium crumenatum*	78
Citrus medica	62	*Dendrobium densiflorum*	78
Citrus reticulata	62	*Dendrocalamus strictus*	79
Clerodendrum serratum	64	*Dendrophthoe falcata*	79
Clerodendrum thomsonae	64	*Desmostachya bipinnata*	80
Clerodendrum viscosum	65	*Dichrostachys cinerea*	80
Clitoria ternatea	65	*Didymosperma nanum*	80
Cochlospermum religiosum	66	*Diplocyclos palmatus*	81
Cocos nucifera	66	*Drypetes roxburghii*	81
Codiaeum variegatum	66	*Elaeagnus latifolia*	83
Coelogyne punctulata	67	*Elaeocarpus ganitrus*	83
Coix gigantea	67	*Elaeocarpus lanceaefolius*	84
Coix lacryma-jobi	67	*Elaeocarpus tectorius*	84
Coix lacryma-jobi var. stenocarpa	68	*Elaeocarpus tuberculatus*	84
		Elettaria cardamomum	85
Commelina benghalensis	68	*Ensete ventricosum*	87

Contd..

Appendix–III–*Contd...*

Contd..

Appendix–III–*Contd...*

Plant Species	*Page No.*	*Plant Species*	*Page No.*
Livistonia speciosa	120	*Pandanus fascicularis*	142
Madhuca indica	122	*Passiflora caerulea*	143
Magnolia campbellii	123	*Pavetta tomentosa*	143
Mallotus philippensis	124	*Phoenix dactylifera*	144
Mangifera indica	124	*Phragmites karka*	144
Manilkara littoralis	125	*Phyllanthus emblica*	145
Melastoma malabathricum	125	*Piper betle*	147
Mesua ferrea	127	*Pogostemon heyneanus*	149
Michelia champaca	128	*Porana paniculata*	150
Mimusops elengi	129	*Potentilla leschenaultiana*	150
Mirabilis jalapa	130	*Pothos scandens*	150
Murraya koenigii	132	*Prunus cerasoides*	151
Murraya paniculata	132	*Prunus mahaleb*	151
Musa paradisiaca	133	*Pterocarpus santalinus*	152
Mussaenda frondosa	133	*Punica roxburghii*	153
Myristica fragrans	134	*Rhynchostylis retusa*	153
Nannorhops ritchieana	134	*Ricinus communis*	154
Naringi crenulata	135	*Rosa clinophylla*	154
Nelumbo nucifera	135	*Rosa indica*	155
Nerium indicum	136	*Rumex scutatus*	156
Nyctanthes arbortristis	137	*Saccharum munja*	156
Nymphaea nouchali	138	*Saccharum officinarum*	157
Nypa fruticans	138	*Saccharum spontaneum*	157
Ocimum basilicum	138	*Samadera indica*	157
Ocimum sanctum	139	*Santalum album*	158
Opilia amentacea	139	*Schima wallichii*	160
Oroxylum indicum	139	*Schrebera sweitenioides*	160
Oryza sativa	140	*Selaginella bisulcata*	161
Osbeckia stellata	140	*Semecarpus anacardium*	162
Oxalis corniculata	141	*Shorea robusta*	162
Pandanus andamanensium	142	*Smithia melenogena var. incanum*	164

Contd..

Appendix–III–*Contd...*

Appendix- IV
Plants Used for Touch Therapy

Contd..

Appendix–IV–*Contd...*

Contd..

Appendix–IV–*Contd...*

Contd..

Appendix–IV–*Contd...*

Appendix-V
Plants Used for Tattooing

Plant Species	*Page No.*	*Plant Species*	*Page No.*
Aegle marmelos	15	*Indigofera dosua*	108
Albizia lebbeck	18	*Lablab purpureus*	116
Aporusa dioica	26	*Oroxylum indicum*	139
Azadirachta indica	32	*Pithecelobium angulatum*	147
Baccaurea sapida	33	*Psychotria calocarpa*	152
Baphicacanthus cusia	33	*Pterocarpus marsupium*	152
Berberis wallichiana	36	*Rubus insignis*	156
Bombax ceiba	38	*Rubus paniculatus*	156
Boswellia serrata	40	*Smilax ocreata*	164
Echinops echinatus	82	*Stereospermum chelonoides*	168
Eclipta prostrata	82	*Taraxacum officinale*	174
Euphorbia prostrarta	91	*Tragopogon dubius*	180

Appendix-VI
Plants Commonly Used for Adornments, Touch Therapy and Tattooing

Plant Species	*Page No.*	*Plant Species*	*Page No.*
Aegle marmelos	15	*Oroxylum indicum*	139
Azadirachta indica	32		

Appendix-VII
Index to Predominant Families

Families	*No. of Species*	*Families*	*No. of Species*
Acanthaceae	8	Lythraceae	4
Amaranthaceae	3	Magnoliaceae	2
Amaryllidaceae	3	Malvaceae	7
Anacardiaceae	4	Melastomataceae	2
Apocynaceae	11	Meliaceae	2
Araceae	5	Menispermaceae	4
Arecaceae	19	Moraceae	5
Asclepiadaceae	5	Musaceae	3
Asteraceae	19	Myrtaceae	3
Balsaminaceae	2	Nyctaginaceae	3
Bignoniaceae	3	Nymphaeaceae	2
Boraginaceae	3	Oleaceae	9
Cannaceae	2	Orchidaceae	7
Capparidaceae	4	Oxalidaceae	3
Celastraceae	3	Pandanaceae	2
Chenopodiaceae	2	Papaveraceae	2
Clusiaceae	3	Pinaceae	2
Combretaceae	7	Plumbaginaceae	2
Convolvulaceae	5	Poaceae	17
Cucurbitaceae	7	Rhamnaceae	2
Dipterocarpaceae	2	Rosaceae	7
Elaeocarpaceae	4	Rubiaceae	15
Euphorbiaceae	20	Rutaceae	9
Fabaceae	49	Sapindaceae	2
Flacourtiaceae	3	Sapotaceae	3
Lamiaceae	6	Schizaeaceae	2
Lauraceae	3	Scrophulariaceae	2
Liliaceae	10	Selaginellaceae	2
Loranthaceae	2	Solanaceae	7

Contd..

Appendix–VII–*Contd...*

Families	*No. of Species*	*Families*	*No. of Species*
Sterculiaceae	3	Ulmaceae	2
Symplocaceae	2	Verbenaceae	12
Tamaricaceae	2	Vitaceae	3
Thymelaeaceae	3	Zingiberaceae	8
Tiliaceae	2		

Appendix-VIII
Index to Predominant Genera of Adornment Plants

Name of Genus	*No. of Species*	*Name of Genus*	*No. of Species*
Acacia	3	*Hibiscus*	2
Aeschynomene	2	*Impatiens*	2
Arnebia	2	*Ipomoea*	2
Bauhinia	2	*Ixora*	2
Borassus	2	*Jasminum*	7
Calamus	4	*Justicia*	2
Callicarpa	2	*Livistonia*	2
Canna	2	*Murraya*	2
Caryota	2	*Ocimum*	2
Citrus	3	*Pandanus*	2
Clerodendrum	3	*Prunus*	2
Coix	3	*Rosa*	2
Crinum	2	*Saccharum*	3
Croton	2	*Symplocos*	2
Curcuma	2	*Tagetes*	2
Dendrobium	2	*Tamarix*	2
Elaeocarpus	4	*Terminalia*	2
Erythrina	2	*Vanda*	2
Euonymus	2	*Zizyphus*	2
Ficus	3		

Appendix-IX
Index to Predominant Genera of Touch Therapy

Name of Genus	*No. of Species*	*Name of Genus*	*No. of Species*
Acacia	2	*Flemingia*	2
Allium	2	*Leucas*	2
Ammannia	2	*Lygodium*	2
Asparagus	2	*Melothria*	2
Biophytum	2	*Musa*	2
Calotropis	2	*Pinus*	2
Cassia	3	*Plumbago*	2
Datura	2	*Solanum*	2
Ficus	2	*Terminalia*	2

Appendix-X
Index to Predominant Genera of Tattooing

Name of Genus	*No. of Species*	*Name of Genus*	*No. of Species*
Aegle	1	*Indigofera*	1
Albizia	1	*Lablab*	1
Aporusa	1	*Oroxylum*	1
Azadirachta	1	*Pithecelobium*	1
Baccaurea	1	*Psychotria*	1
Baphicacanthus	1	*Pterocarpus*	1
Berberis	1	*Rubus*	2
Bombax	1	*Smilax*	1
Boswellia	1	*Stereospermum*	1
Echinops	1	*Taraxacum*	1
Eclipta	1	*Tragopogon*	1
Euphorbia	1		

Appendix-XI
Index to Plant Parts Used for Adornments, Touch Therapy and Tattooing

Part/s Used	*Plant*
Aerial parts	*Actinopteris radiata, Artemisia nilagirica, Asplenium aethiopium, Bulbophyllum neilgherrense, Rumex scutatus.*
Apical buds	*Tectona grandis.*
Arils	*Myristica fragrans.*
Beads	*Cascabela thevetia, Casearia graveolens, Erythrina indica, Euonymus grandiflorum.*
Bracts	*Coix lacryma-jobi.*
Branches	*Clerodendrum indicum, Lablab purpureus, Trichodesma indicum.*
Bulb	*Allium cepa, Allium sativum.*
Culms	*Bambusa vulgaris, Phragmites karka, Triticum aestivum.*
Fibre	*Agave americana, Cocos nucifera, Gnetum scandens, Hibiscus tiliaceus.*
Flower buds	*Smithia melenogena* var. *incanum, Solanum melongena* var. *incarum.*
Flowers	*Acacia ferruginea, Achilla millefolium, Aegiceras corniculatum, Albizzia lebbeck, Alpinia allughas, Anthocephalus chinensis, Barleria prionitis, Basella alba, Bauhinia racemosa, Bougainvillaea spectabilis, Butea monosperma, Calophyllum inophyllum, Calotropis gigantea, Canna speciosa, Carisa spinarum, Cascabela thevetia, Cassia auriculata, Catharanthus roseus, Celosia argentea, Citrus medica, Clerodendrum thomsonae, Clerodendrum viscosum, Clitoria ternatea, Cochlospermum religiosum, Coelogyne punctulata, Commelina benghalensis, Crinum asiaticum, Crinum latifolium, Crossandra undulaefolia, Curcuma neilgherrensis, Curcuma zedoaria, Daphne papyracea, Delonix regia, Ervatamia heyneana, Erythrina indica, Erythrina suberosa, Eugenia caryophyllata, Garcinia morella, Geranium himalayense, Globba marantina, Gloriosa superba, Guettarda speciosa, Helianthus annuus, Hibiscus rosa-sinensis, Hiptage benghalensis, Holarrhena antidysenterica, Hornstedtia arunachalensis, Ipomoea aquatica, Ipomoea pes-caprae, Ixora barbata, Ixora brunnescens, Jacaranda mimosifolia, Jasminum angustifolium, Jasminum arborescens, Jasminum auriculatum, Jasminum flexile, Jasminum grandiflorum, Jasminum multiflorum, Jasminum sambac, Lawsonia alba, Leucas aspera, Madhuca indica, Magnolia campbellii, Manilkara littoralis, Melastoma malabathricum, Mesua ferrea, Michelia champaca, Mimusops elengi, Mirabilis jalapa, Murraya paniculata, Nelumbo nucifera, Nerium indicum, Nyctanthes arbortristis, Nymphaea nouchali, Opilia amentacea, Oxalis corniculata,*

Contd..

Appendix–XI–*Contd...*

Part/s Used	*Plant*
	Pandanus fascicularis, Passiflora caerulea, Pavetta tomentosa, Pogostemon heyneanus, Porana paniculata, Rhynchostylis retusa, Rosa clinophylla, Rosa indica, Saraca asoca, Schima wallichii, Shorea robusta, Smilax ocreata, Spermadictyon sauveolens, Sphaeranthus indicus, Syzygium jambos, Tabernaemontana divaricata, Tadehagi triquetrum, Tagetes erecta, Tagetes patula, Tridax procumbens, Trigonella corniculata, Vanda teres, Wrightia tinctoria, Xanthium strumarium.
Fruits	*Agaricus burkillii, Baccaurea courtallensis, Borassus ramiflora, Calotropis procera, Capsicum annuum, Carica papaya, Cassia fistula, Celastrus paniculatus, Citrus maxima, Citrus medica, Citrus reticulate, Cocos nucifera, Coix gigantean, Coix lacryma-jobi* var. *stenocarpa, Crotalaria spectabilis, Drypetes roxburghii, Elaeagnus latifolia, Ervatamia heyneana, Gossypium arboretum, Guazuma ulmifolia, Helicteres isora, Indigofera cassioides, Lagenaria siceraria, Leucaena leucocephala, Mallotus philippensis, Melia azedarach, Melothria maderaspatana, Oroxylum indicum, Phyllanthus emblica, Pongamia pinnata, Schrebera sweitenioides, Semecarpus anacardium, Smilax ocreata, Solanum anguivi, Solanum surattense, Spondias pinnata, Stereospermum chelonoides, Symphorema polyandrum, Syzygium jambos, Thunbergia fragrans, Tribulus terrestris, Vitex negundo, Xanthium strumarium, Xeromphis spinosa, Zizyphus mauritiana.*
Girdles	*Saccharum munja.*
Inflorescence	*Phragmites karka, Saccharum officinarum, Saccharum spontaneum, Tamarix aphylla.*
Kernels	*Cocos nucifera, Prunus mahaleb.*
Latex	*Calotropis gigantea, Euphorbia prostrarta, Hevea brasiliensis, Taraxacum officinale, Tragopogon dubius.*
Leaves	*Achyranthes aspera, Ageratum conyzoides, Aloe barbadensis, Argyreia nervosa, Artemisia nilagirica, Asparagus racemosus, Azadirachta indica, Baphicacanthus cusia, Begonia picta, Bergenia ciliata, Borassus flabellifer, Bryophyllum pinnatum, Caladium bicolor, Callicarpa arborea, Callicarpa macrophylla, Calotropis gigantea, Calotropis procera, Cannabis sativa, Caryota mitis, Casearia elliptica, Cassia tora, Chenopodium album, Citrus medica, Claoxylon indicum, Cleome viscosa, Clerodendrum serratum, Cocos nucifera, Codiaeum variegatum, Crataeva magna, Curculigo capitulata, Daphne papyracea, Datura fastuosa* var. *alba, Datura innoxia, Eclipta prostrata, Enicostemma axillare, Eucalyptus maculata, Ficus racemosa, Ficus religiosa, Haldina cordifolia, Holarrhena antidysenterica, Impatiens balsamina, Indigofera dosua, Justicia gendarussa, Lablab purpureus, Launaea procumbens,*

Contd..

Appendix–XI–*Contd...*

Part/s Used	*Plant*
	Lawsonia alba, Licuala peltata, Livistonia jenkinsiana, Livistonia speciosa, Mallotus peltatus, Mangifera indica, Manilkara littoralis, Mesua ferrea, Mimosa pudica, Molineria prainiana, Momordica charantia, Musa paradisiaca, Mussaenda frondosa, Naravelia zeylanica, Naringi crenulata, Neolitsea chinensis, Nicotiana tabaccum, Nypa fruticans, Ocimum sanctum, Osbeckia stellata, Pandanus andamanensium, Pandanus fascicularis, Papaver somniferum, Phoenix dactylifera, Phoenix sylvestris, Pinus kesiya, Piper betle, Pogostemon heyneanus, Pongamia pinnata, Pothos scandens, Psychotria calocarpa, Quisqualis indica, Ricinus communis, Saccharum officinarum, Selaginella bisulcata, Tabernaemontana divaricata, Tamarix dioica, Taxus baccata, Terminalia bialata, Terminalia tomentosa, Thysanolaena maxima, Vanda tessellate, Vitex negundo, Wattakaka volubilis.
Nuts	*Areca catechu, Elaeocarpus ganitrus, Elaeocarpus lanceaefolius, Elaeocarpus tuberculatus.*
Pith	*Aeschynomene aspera, Aeschynomene indica, Hymenachne amplexicaulis.*
Pollen	*Symplocos racemosa.*
Resins	*Pinus wallichiana, Vateria indica.*
Rhizomes	*Acorus calamus, Curcuma longa.*
Roots	*Abelmoschus moschatus, Abrus precatorius, Acacia nilotica, Acacia pennata, Achyranthes aspera, Acorus calamus, Adiantum caudatum, Aegle marmelos, Alternanthera sessilis, Alysicarpus vaginalis, Antidesma diandrum, Argemone mexicana, Arnebia benthamii, Arnebia guttata, Asparagus gonocladus, Asparagus racemosus, Boerhaavia diffusa, Borreria articularis, Butea monosperma, Callicarpa macrophylla, Calotropis procera, Capparis sepiaria, Capsicum annuum, Cassia tora, Ceriscoides turgida, Clerodendrum serratum, Clitoria ternatea, Corchorus capsularis, Croton roxburghii, Datura fastuosa* var. *alba, Datura innoxia, Dendrocalamus strictus, Diplocyclos palmatus, Echinops echinatus, Elephantopus scaber, Elytraria acaulis, Euphorbia ligularia, Ficus racemosa, Flemingia bracteata, Gardenia turgida, Gloriosa superba, Grewia subinaequalis, Helminthostachys zeylanica, Hemidesmus indicus, Holarrhena antidysenterica, Ichnocarpus frutescens, Indigofera astragalina, Jatropha gossypifolia, Kaempferia angustifolia, Launaea procumbens, Leucas aspera, Luffa acutangula, Lugodium flexuosum, Melothria heterophylla, Mimosa pudica, Mirabilis jalapa, Mucuna pruriens, Musa paradisiacal, Ocimum sanctum, Phoenix sylvestris, Plumbago indica, Plumbago zeylanica, Pongamia pinnata, Potentilla leschenaultiana, Premna serratifolia, Ricinus communis, Rubia*

Contd..

Appendix–XI–*Contd...*

Part/s Used	*Plant*
	cordifolia, Scoparia dulcis, Securinega virosa, Sida rhombifolia, Smilax wightii, Tephrosia purpurea, Terminalia arjuna, Thelypteris extensa, Thysanolaena maxima, Urena lobata, Urtica dioica, Vernonia cinerea, Vetiveria zizanioides, Vitex negundo, Woodfordia fruticosa.
Seeds	*Abrus precatorius, Abutilon indicum, Adenanthera pavonina, Aleurites moluccana, Anadendrum paniculatum, Anisophyllea zeylanica, Bixa orellana, Blastania garcinii, Brassica alba, Caesalpinia bonduc, Calotropis gigantean, Canna orientalis, Cardiospermum halicacabum, Caryota urens, Coix lacryma-jobi, Corypha umbraculifera, Dalbergia sissoo, Drypetes roxburghii, Elaeocarpus ganitrus, Elaeocarpus tectorius, Elettaria cardamomum, Ensete ventricosum, Entada phaseoloides, Euonymus fimbriatus, Flueggea leucopyrus, Gyrocarpus americanus, Kaempferia angustifolia, Lablab purpureus, Lannea caromandelica, Leea macrophylla, Mimusops elengi, Mirabilis jalapa, Murraya koenigii, Musa glauca, Musa paradisiaca, Nannorhops ritchieana, Nelumbo nucifera, Nymphaea nouchali, Oroxylum indicum, Oryza sativa, Ricinus communis, Samadera indica, Semecarpus anacardium, Stereospermum chelonoides, Symplocos cochinchinensis, Tamarindus indica, Taxus baccata, Terminalia chebula, Thysanolaena maxima, Xanthium strumarium, Zizyphus xylopyrus.*
Shoots	*Caladium bicolor, Cannabis sativa, Cassia auriculata, Codiaeum variegatum, Lablab purpureus, Livistonia jenkinsiana, Phragmites karka, Saccharum officinarum, Selaginella bisulcata, Tamarix dioica.*
Stem	*Abrus precatorius, Achyranthes aspera, Aeschynomene aspera, Ajuga macrosperma, Alpinia allughas, Ampelocissus latifolia, Atylosia volubilis, Bauhinia vahlii, Calamus erectus, Calamus floribundus, Calamus inermis, Calamus latifolia, Careya arborea, Cissampelos pareira, Cissus quadrangularis, Congea tomentosa, Croton caudatus, Dendrobium crumenatum, Dendrocalamus strictus, Dendrophthoe falcata,Diplocyclos palmatus, Equisetum debile, Euphorbia ligularia, Gloriosa superba, Gnidia glauca, Grewia subinaequalis, Holoptelea integrifolia, Ichnocarpus frutescens, Ipomoea pes caprae, Leucas cephalotes, Linum usitatissimum, Mimosa pudica, Musa paradisiaca, Naravelia zeylanica, Ocimum basilicum, Ocimum sanctum, Paederia foetida, Phyla nodiflora, Phyllocephalum phyllolaenum, Pothos scandens, Shuteria densiflora, Sphaeranthus indicus, Thunbergia fragrans, Tinospora cordifolia, Toddalia asiatica.*
Stem bark	*Abroma augusta, Acacia leucophloea, Aegle marmelos, Albizzia lebbeck, Annona squamosa, Aporusa dioica, Baccaurea sapida, Bauhinia racemosa, Bombax ceiba, Boswellia serrata, Butea monosperma, Callicarpa arborea, Calotropis procera, Cassia fistula, Cayratia pedata, Celtis cinnamomea,*

Contd..

Appendix–XI–*Contd...*

Part/s Used	*Plant*
	Cordia gharaf, Dendrobium densiflorum, Dichrostachys cinerea, Erythrina indica, Ficus altissima, Ficus hispida, Ficus racemosa, Holoptelea integrifolia, Juglans regia, Litsea monopetala, Lobelia nicotianifolia, Manilkara littoralis, Nyctanthes arbortristis, Oroxylum indicum, Pinus wallichiana, Pithecelobium angulatum, Sapindus emarginata, Wrightia arborea.
Stones	*Azadirachta indica, Prunus cerasoides, Punica roxburghii.*
Tendrils	*Bauhinia racemosa, Bauhinia vahlii.*
Thorns	*Aegle marmelos, Berberis wallichiana, Cirsium lipskyi, Rubus insignis, Rubus paniculatus.*
Tuber	*Gloriosa superba, Stephania glabra, Tuber aestivum.*
Twigs	*Acacia pennata, Alstonia scholaris, Asparagus racemosus, Azadirachta indica, Euclea fruiticosa, Flemingia chappar, Phyllanthus emblica, Streblus asper, Taxus baccata, Uraria picta.*
Whole Plant	*Amanita muscaria, Ammannia pentandra, Ammannia peploides, Artemisia nilagirica, Biophytum apodiscias, Biophytum reinwardtii, Capparis aphylla, Cassytha filiformis, Caulerpa racemosa, Chlorophytum arundinaceum, Clitoria ternatea, Cynodon dactylon, Dendrophthoe falcata, Desmostachya bipinnata, Didymosperma nanum, Emilia sonchifolia, Evolvulus nummularius, Hemidesmus indicus, Hordeum vulgare, Hydrocotyle javanica, Impatiens gigantea, Justicia betonica, Lygodium japonicum, Marsilea minuta, Millettia fruticosa, Mimosa pudica, Pterocarpus marsupium, Scurrula parasitica, Selaginella brachystachya, Stephania hernandifolia, Striga gesnerioides, Tylophora indica, Vernonia cinerea.*
Wood	*Acacia catechu, Aegle marmelos, Anogeissus latifolia, Aquilaria agallocha, Cajanus cajan, Cassia fistula, Crataeva magna, Erythrina indica, Flacourtia ramontchi, Justicia adhatoda, Madhuca indica, Pterocarpus santalinus, Sambucus nigra, Santalum album, Tamarix aphylla, Zanthoxylum limonella, Zizyphus mauritiana.*

Appendix-XII
Index to Plant Parts Used for Adornments

Part Used	*No. of Species*	*Plant Species*
Flowers	94	*Acacia ferruginea, Achilla millefolium, Aegiceras corniculatum, Albizzia lebbeck, Alpinia allughas, Anthocephalus chinensis, Barleria prionitis, Basella alba, Bauhinia racemosa, Bougainvillaea spectabilis, Butea monosperma, Calophyllum inophyllum, Calotropis gigantea, Canna speciosa, Carisa spinarum, Cascabela thevetia, Catharanthus roseus, Celosia argentea, Citrus medica, Clerodendrum thomsonae, Clitoria ternatea, Cochlospermum religiosum, Coelogyne punctulata, Commelina benghalensis, Crinum asiaticum, Crinum latifolium, Crossandra undulaefolia, Curcuma neilgherrensis, Curcuma zedoaria, Daphne papyracea, Delonix regia, Ervatamia heyneana, Erythrina indica, Erythrina suberosa, Eugenia caryophyllata, Garcinia morella, Geranium himalayense, Globba marantina, Gloriosa superba, Guettarda speciosa, Helianthus annuus, Hibiscus rosa-sinensis, Hiptage benghalensis, Holarrhena antidysenterica, Hornstedtia arunachalensis, Ipomoea aquatica, Ipomoea pes-caprae, Ixora barbata, Ixora brunnescens, Jacaranda mimosifolia, Jasminum angustifolium, Jasminum arborescens, Jasminum auriculatum, Jasminum flexile, Jasminum grandiflorum, Jasminum multiflorum, Jasminum sambac, Lawsonia alba, Leucas aspera, Madhuca indica, Magnolia campbellii, Manilkara littoralis, Melastoma malabathricum, Mesua ferrea, Michelia champaca, Mimusops elengi, Mirabilis jalapa, Murraya paniculata, Nelumbo nucifera, Nerium indicum, Nyctanthes arbortristis, Nymphaea nouchali, Opilia amentacea, Oxalis corniculata, Pandanus fascicularis, Passiflora caerulea, Pavetta tomentosa, Pogostemon heyneanus, Porana paniculata, Rhynchostylis retusa, Rosa clinophylla, Rosa indica, Schima wallichii, Shorea robusta, Spermadictyon sauveolens, Sphaeranthus indicus, Syzygium jambos, Tabernaemontana divaricata, Tadehagi triquetrum, Tagetes erecta, Tagetes patula, Trigonella corniculata, Vanda teres, Wrightia tinctoria.*
Leaves	44	*Asparagus racemosus, Begonia picta, Borassus flabellifer, Caladium bicolor, Callicarpa arborea, Callicarpa macrophylla, Calotropis gigantea, Cannabis sativa, Caryota mitis, Citrus medica, Cocos nucifera, Codiaeum variegatum, Daphne papyracea, Eucalyptus maculata, Ficus racemosa, Haldina cordifolia, Holarrhena antidysenterica, Impatiens balsamina,*

Contd..

Appendix–XII–*Contd...*

Part Used	*No. of Species*	*Plant Species*
		Lawsonia alba, Licuala peltata, Livistonia jenkinsiana, Livistonia speciosa, Mangifera indica, Manilkara littoralis, Mesua ferrea, Musa paradisiaca, Mussaenda frondosa, Naringi crenulata, Nypa fruticans, Ocimum sanctum, Osbeckia stellata, Pandanus andamanensium, Pandanus fascicularis, Phoenix dactylifera, Piper betle, Pogostemon heyneanus, Pothos scandens, Saccharum officinarum, Selaginella bisulcata, Tabernaemontana divaricata, Tamarix dioica, Terminalia bialata, Terminalia tomentosa, Vanda tessellata.
Seeds	39	*Abrus precatorius, Abutilon indicum, Adenanthera pavonina, Aleurites moluccana, Anadendrum paniculatum, Anisophyllea zeylanica, Bixa orellana, Blastania garcinii, Caesalpinia bonduc, Canna orientalis, Caryota urens, Coix lacryma-jobi, Corypha umbraculifera, Dalbergia sissoo, Elaeocarpus tectorius, Elettaria cardamomum, Ensete ventricosum, Entada phaseoloides, Euonymus fimbriatus, Flueggea leucopyrus, Gyrocarpus americanus, Kaempferia angustifolia, Lannea caromandelica, Mimusops elengi, Mirabilis jalapa, Murraya koenigii, Musa paradisiaca, Nannorhops ritchieana, Nelumbo nucifera, Nymphaea nouchali, Oroxylum indicum, Oryza sativa, Ricinus communis, Samadera indica, Semecarpus anacardium, Symplocos cochinchinensis, Taxus baccata, Thysanolaena maxima, Zizyphus xylopyrus.*
Fruit	23	*Baccaurea courtallensis, Borassus ramiflora, Capsicum annuum, Celastrus paniculatus, Citrus maxima, Citrus medica, Citrus reticulate, Cocos nucifera, Coix gigantean, Coix lacryma-jobi* var. *stenocarpa, Crotalaria spectabilis, Drypetes roxburghii, Gossypium arboretum, Guazuma ulmifolia, Lagenaria siceraria, Leucaena leucocephala, Mallotus philippensis, Phyllanthus emblica, Schrebera sweitenioides, Syzygium jambos, Thunbergia fragrans, Xeromphis spinosa, Zizyphus mauritiana.*
Roots	17	*Acacia pennata, Achyranthes aspera, Arnebia benthamii, Arnebia guttata, Callicarpa macrophylla, Ceriscoides turgida, Clerodendrum serratum, Croton roxburghii, Diplocyclos palmatus, Gardenia turgida, Hemidesmus indicus, Kaempferia angustifolia, Launaea procumbens, Ocimum sanctum, Potentilla leschenaultiana, Ricinus communis, Urtica dioica.*
Stem	17	*Aeschynomene aspera, Ajuga macrosperma, Bauhinia vahlii, Calamus erectus, Calamus floribundus, Calamus inermis,*

Contd..

Appendix–XII–*Contd...*

Part Used	*No. of Species*	*Plant Species*
		Calamus latifolia, Cissus quadrangularis, Croton caudatus, Dendrobium crumenatum, Dendrocalamus strictus, Dendrophthoe falcata, Ipomoea pes caprae, Linum usitatissimum, Ocimum basilicum, Ocimum sanctum, Pothos scandens.
Wood	15	*Acacia catechu, Aegle marmelos, Anogeissus latifolia, Aquilaria agallocha, Cajanus cajan, Crataeva magna, Flacourtia ramontchi, Justicia adhatoda, Madhuca indica, Pterocarpus santalinus, Sambucus nigra, Santalum album, Tamarix aphylla, Zanthoxylum limonella, Zizyphus mauritiana.*
Whole plant	13	*Amanita muscaria, Ammannia pentandra, Ammannia peploides, Capparis aphylla, Caulerpa racemosa, Cynodon dactylon, Desmostachya bipinnata, Didymosperma nanum, Hordeum vulgare, Impatiens gigantea, Justicia betonica,Stephania hernandifolia, Striga gesnerioides.*
Stem bark	10	*Callicarpa arborea, Celtis cinnamomea, Cordia gharaf, Dendrobium densiflorum, Dichrostachys cinerea, Erythrina indica, Ficus altissima, Ficus hispida, Ficus racemosa, Nyctanthes arbortristis.*
Shoots	7	*Caladium bicolor, Cannabis sativa, Codiaeum variegatum, Livistonia jenkinsiana, Saccharum officinarum, Selaginella bisulcata, Tamarix dioica.*
Twigs	5	*Acacia pennata, Alstonia scholaris, Asparagus racemosus, Euclea fruiticosa, Taxus baccata.*
Aerial parts	4	*Actinopteris radiata, Asplenium aethiopium, Bulbophyllum neilgherrense, Rumex scutatus.*
Beads	4	*Cascabela thevetia, Casearia graveolens, Erythrina indica, Euonymus grandiflorum.*
Fibre	4	*Agave americana, Cocos nucifera, Gnetum scandens, Hibiscus tiliaceus.*
Nuts	4	*Areca catechu, Elaeocarpus ganitrus, Elaeocarpus lanceaefolius, Elaeocarpus tuberculatus.*
Culms	3	*Bambusa vulgaris, Phragmites karka, Triticum aestivum.*
Inflorescence	3	*Saccharum officinarum, Saccharum spontaneum, Tamarix aphylla.*
Pith	3	*Aeschynomene aspera, Aeschynomene indica, Hymenachne amplexicaulis.*

Contd..

Appendix–XII–*Contd...*

Part Used	*No. of Species*	*Plant Species*
Stones	3	*Azadirachta indica, Prunus cerasoides, Punica roxburghii.*
Flower buds	2	*Smithia melenogena* var. *incanum, Solanum melongena* var. *incarnum.*
Kernels	2	*Cocos nucifera, Prunus mahaleb.*
Latex	2	*Calotropis gigantea, Hevea brasiliensis.*
Tendrils	2	*Bauhinia racemosa, Bauhinia vahlii.*
Apical buds	1	*Tectona grandis.*
Arils	1	*Myristica fragrans.*
Bracts	1	*Coix lacryma-jobi.*
Branches	1	*Clerodendrum indicum.*
Girdles	1	*Saccharum munja.*
Pollen	1	*Symplocos racemosa.*
Resins	1	*Vateria indica.*
Tuber	1	*Tuber aestivum.*

Appendix-XIII
Index to Plant Parts Used for Touch Therapy

Part Used	*No. of Species*	*Plant Species*
Roots	71	*Abelmoschus moschatus, Abrus precatorius, Acacia leucophloea, Acacia nilotica, Achyranthes* aspera, *Acorus calamus, Adiantum caudatum, Aegle marmelos, Alternanthera sessilis, Alysicarpus vaginalis, Antidesma diandrum, Argemone mexicana, Asparagus gonocladus, Asparagus racemosus, Boerhaavia diffusa, Borreria articularis, Butea monosperma, Calotropis procera, Capparis sepiaria, Capsicum annuum, Cassia tora, Cayratia pedata, Clerodendrum serratum, Clitoria ternatea, Corchorus capsularis, Datura fastuosa, Datura innoxia, Dendrocalamus strictus, Diplocyclos palmatus, Echinops echinatus, Elephantopus scaber, Elytraria acaulis, Euphorbia ligularia, Ficus racemosa, Flemingia bracteata, Gloriosa superba, Grewia subinaequalis, Helminthostachys zeylanica, Hemidesmus indicus, Holarrhena antidysenterica, Ichnocarpus frutescens, Indigofera astragalina, Jatropha gossypifolia, Leucas aspera, Luffa acutangula, Lygodium flexuosum, Melothria heterophylla, Mimosa pudica, Mirabilis jalapa, Mucuna pruriens, Musa paradisiaca, Phoenix sylvestris, Plumbago indica, Plumbago zeylanica, Pongamia pinnata, Premna serratifolia, Rubia cordifolia, Scoparia dulcis, Securinega virosa, Sida rhombifolia, Smilax wightii, Tephrosia purpurea, Terminalia arjuna, Thelypteris extensa, Thysanolaena maxima, Urena lobata, Vernonia cinerea, Vetiveria zizanioides, Vitex negundo, Woodfordia fruticosa.*
Leaves	43	*Achyranthes aspera, Ageratum conyzoides, Aloe barbadensis, Argyreia nervosa, Artemisia nilagirica, Azadirachta indica, Bergenia ciliata, Bryophyllum pinnatum, Calotropis procera, Casearia elliptica, Cassia tora, Chenopodium album, Claoxylon indicum, Cleome viscosa, Clerodendrum serratum, Crataeva magna, Curculigo capitulata, Datura fastuosa, Datura innoxia, Enicostemma axillare, Ficus religiosa, Justicia gendarussa, Launaea procumbens, Luffa acutangula, Mallotus peltatus, Mimosa pudica, Molineria prainiana, Momordica charantia, Naravelia zeylanica, Neolitsea chinensis, Nicotiana tabaccum, Papaver somniferum, Phoenix sylvestris, Pinus kesiya, Piper betle, Pongamia pinnata, Quisqualis indica, Ricinus communis, Tabernaemontana divaricata, Taxus baccata, Thysanolaena maxima, Vitex negundo, Wattakaka volubilis.*

Contd..

Appendix–XIII–*Contd...*

Part Used	*No. of Species*	*Plant Species*
Stem	31	*Abrus precatorius, Achyranthes aspera, Ajuga macrosperma, Alpinia allughas, Ampelocissus latifolia, Atylosia volubilis, Careya arborea, Cissampelos pareira, Cissus quadrangularis, Congea tomentosa, Dendrocalamus strictus, Dendrophthoe falcata, Diplocyclos palmatus, Equisetum debile, Euphorbia ligularia, Gloriosa superba, Gnidia glauca, Grewia subinaequalis, Ichnocarpus frutescens, Leucas cephalotes, Mimosa pudica, Musa paradisiaca, Naravelia zeylanica, Paederia foetida, Phyla nodiflora, Phyllocephalum phyllolaenum, Shuteria densiflora, Sphaeranthus indicus, Thunbergia fragrans, Tinospora cordifolia, Toddalia asiatica.*
Whole plant	21	*Ammania pentandra, Ammannia peploides, Artemisia nilagirica, Biophytum apodiscias,Biophytum reinwardtii, Cassytha filiformis, Chlorophytum arundinaceum, Clitoria ternatea, Dendrophthoe falcata, Emilia sonchifolia, Evolvulus nummularius, Hemidesmus indicus, Hydrocotyle javanica, Lygodium japonicum, Marsilea minuta, Millettia fruticosa, Mimosa pudica, Scurrula parasitica, Selaginella brachystachya, Tylophora indica, Vernonia cinerea.*
Fruit	20	*Agaricus burkillii, Calotropis procera, Carica papaya, Cassia fistula, Drypetes roxburghii, Ervatamia heyneana, Helicteres isora, Melia azedarach, Melothria maderaspatana, Oroxylum indicum, Pongamia pinnata, Semecarpus anacardium, Solanum anguivi, Solanum surattense, Spondias pinnata, Stereospermum chelonoides, Symphorema polyandrum, Tribulus terrestris, Vitex negundo, Xanthium strumarium.*
Stem bark	16	*Abroma augusta, Acacia leucophloea, Aegle marmelos, Annona squamosa, Bauhinia racemosa, Bombax ceiba, Butea monosperma, Calotropis procera, Cassia fistula, Holoptelea integrifolia, Litsea monopetala, Lobelia nicotianifolia, Manilkara littoralis, Pinus wallichiana, Sapindus emarginata, Wrightia arborea.*
Seeds	14	*Abrus precatorius, Abutilon indicum, Brassica alba, Caesalpinia bonduc, Calotropis gigantea, Cardiospermum halicacabum, Elaeocarpus ganitrus, Ficus racemosa, Leea macrophylla, Musa glauca, Stereospermum chelonoides, Tamarindus indica, Terminalia chebula, Xanthium strumarium.*
Flowers	5	*Cassia auriculata, Saraca asoca, Sphaeranthus indicus, Tridax procumbens, Xanthium strumarium.*

Contd..

Appendix–XIII–*Contd...*

Part Used	*No. of Species*	*Plant Species*
Twigs	5	*Azadirachta indica, Flemingia chappar, Phyllanthus emblica, Streblus asper, Uraria picta.*
Wood	4	*Aegle marmelos, Cassia fistula, Erythrina indica, Sambucus nigra.*
Bulb	2	*Allium cepa, Allium sativum.*
Rhizomes	2	*Acorus calamus, Curcuma longa.*
Shoots	2	*Cassia auriculata, Phragmites karka.*
Tuber	2	*Gloriosa superba, Stephania glabra.*

Appendix-XIV
Index to Plant Parts Used for Tattooing

Part Used	*No. of Species*	*Plant Species*
Stem bark	7	*Albizia lebbeck, Aporusa dioica, Baccaurea sapida, Bombax ceiba,Boswellia serrata,Oroxylum indicum, Pithecellobium angulatum.*
Leaves	6	*Azadirachta indica, Baphicacanthus cusia, Eclipta prostrata, Indigofera dosua, Lablab purpureus, Psychotria calocarpa.*
Thorns	4	*Aegle marmelos, Berberis wallichiana, Rubus insignis, Rubus paniculatus.*
Latex	3	*Euphorbia prostrarta, Taraxacum officinale, Tragopogon dubius.*
Fruits	2	*Smilax ocreata, Stereospermum chelonoides*
Flowers	1	*Smilax ocreata.*
Roots	1	*Echinops echinatus.*
Whole plant	1	*Pterocarpus marsupium.*

Appendix-XV
Index to Predominant Dicot Families of Plants Used for Adornments, Touch Therapy and Tattooing with Relative per cent age in Taxa

Families	*Total No. of Species*	*Per cent in Taxa*
Acanthaceae	8	1.81 per cent
Amaranthaceae	3	0.68 per cent
Anacardiaceae	4	0.90 per cent
Apocynaceae	11	2.50 per cent
Asclepiadaceae	5	1.13 per cent
Asteraceae	19	4.31 per cent
Balsaminaceae	2	0.45 per cent
Bignoniaceae	3	0.68 per cent
Boraginaceae	3	0.68 per cent
Capparidaceae	4	0.90 per cent
Celastraceae	3	0.68 per cent
Chenopodiaceae	2	0.45 per cent
Clusiaceae	3	0.68 per cent
Combretaceae	7	1.59 per cent
Convolvulaceae	5	1.13 per cent
Cucurbitaceae	7	1.59 per cent
Dipterocarpaceae	2	0.45 per cent
Elaeocarpaceae	4	0.90 per cent
Euphorbiaceae	20	4.54 per cent
Fabaceae	49	11.13 per cent
Flacourtiaceae	3	0.68 per cent
Lamiaceae	6	1.36 per cent
Lauraceae	3	0.68 per cent
Loranthaceae	2	0.45 per cent
Lythraceae	4	0.90 per cent
Magnoliaceae	2	0.45 per cent
Malvaceae	7	1.59 per cent
Melastomataceae	2	0.45 per cent

Contd...

Appendix–XV–*Contd...*

Families	*Total No. of Species*	*Per cent in Taxa*
Meliaceae	2	0.45 per cent
Menispermaceae	4	0.90 per cent
Moraceae	5	1.13 per cent
Myrtaceae	3	0.68 per cent
Nyctaginaceae	3	0.68 per cent
Nymphaeaceae	2	0.45 per cent
Oleaceae	9	2.04 per cent
Oxalidaceae	3	0.68 per cent
Papaveraceae	2	0.45 per cent
Plumbaginaceae	2	0.45 per cent
Rhamnaceae	2	0.45 per cent
Rosaceae	7	1.59 per cent
Rubiaceae	15	3.40 per cent
Rutaceae	9	2.04 per cent
Sapindaceae	2	0.45 per cent
Sapotaceae	3	0.68 per cent
Scrophulariaceae	2	0.45 per cent
Solanaceae	7	1.59 per cent
Sterculiaceae	3	0.68 per cent
Symplocaceae	2	0.45 per cent
Tamaricaceae	2	0.45 per cent
Thymelaeaceae	3	0.68 per cent
Tiliaceae	2	0.45 per cent
Ulmaceae	2	0.45 per cent
Verbenaceae	12	2.72 per cent
Vitaceae	3	0.68 per cent

Appendix-XVI
Index to Predominant Monocot Families of Plants used for Adornments, Touch Therapy and Tattooing

Families	*Total No. of Species*	*Per cent in Taxa*
Arecaceae	19	4.31 per cent
Poaceae	17	3.86 per cent
Liliaceae	10	2.27 per cent
Zingiberaceae	8	1.81 per cent
Orchidaceae	7	1.59 per cent
Araceae	5	1.13 per cent
Amaryllidaceae	3	0.68 per cent
Musaceae	3	0.68 per cent
Cannaceae	2	0.45 per cent
Pandanaceae	2	0.45 per cent

Appendix-XVII
Index to Botanical Names

Contd..

Appendix–XVII–*Contd...*

Contd..

Appendix–XVII–*Contd...*

Plant Species	*Page No.*	*Plant Species*	*Page No.*
Chlorophytum arundinaceum	59	*Crotalaria spectabilis*	72
Cirsium lipskyi	60	*Croton caudatus*	72
Cissampelos pareira	60	*Croton roxburghii*	73
Cissus quadrangularis	61	*Curculigo capitulata*	73
Citrus maxima	61	*Curcuma longa*	73
Citrus medica	62	*Curcuma neilgherrensis*	74
Citrus reticulata	62	*Curcuma zedoaria*	74
Claoxylon indicum	63	*Cynodon dactylon*	75
Cleome viscosa	63	*Dalbergia sissoo*	75
Clerodendrum indicum	63	*Daphne papyracea*	76
Clerodendrum serratum	64	*Datura fastuosa* var. *alba*	76
Clerodendrum thomsonae	64	*Datura innoxia*	77
Clerodendrum viscosum	65	*Delonix regia*	77
Clitoria ternatea	65	*Dendrobium crumenatum*	78
Cochlospermum religiosum	66	*Dendrobium densiflorum*	78
Cocos nucifera	66	*Dendrocalamus strictus*	79
Codiaeum variegatum	66	*Dendrophthoe falcata*	79
Coelogyne punctulata	67	*Desmostachya bipinnata*	80
Coix gigantea	67	*Dichrostachys cinerea*	80
Coix lacryma-jobi	67	*Didymosperma nanum*	80
Coix lacryma-jobi var. *stenocarpa*	68	*Diplocyclos palmatus*	81
		Drypetes roxburghii	81
Commelina benghalensis	68	*Echinops echinatus*	82
Congea tomentosa	69	*Eclipta prostrata*	82
Corchorus capsularis	69	*Elaeagnus latifolia*	83
Cordia gharaf	69	*Elaeocarpus ganitrus*	83
Corypha umbraculifera	70	*Elaeocarpus lanceaefolius*	84
Crataeva magna	70	*Elaeocarpus tectorius*	84
Crinum asiaticum	71	*Elaeocarpus tuberculatus*	84
Crinum latifolium	71	*Elephantopus scaber*	84
Crossandra undulaefolia	72	*Elettaria cardamomum*	85

Contd..

Appendix–XVII–*Contd...*

Plant Species	*Page No.*	*Plant Species*	*Page No.*
Elytraria acaulis	85	*Gnidia glauca*	98
Emilia sonchifolia	86	*Gossypium arboreum*	98
Enicostema axillare	86	*Grewia subinaequalis*	98
Ensete ventricosum	87	*Guazuma ulmifolia*	99
Entada phaseoloides	87	*Guettarda speciosa*	99
Equisetum debile	87	*Gyrocarpus americanus*	99
Ervatamia heyneana	88	*Haldina cordifolia*	100
Erythrina indica	88	*Helianthus annuus*	100
Erythrina suberosa	89	*Helicteres isora*	101
Eucalyptus maculata	89	*Helminthostachys zeylanica*	101
Euclea fruiticosa	90	*Hemidesmus indicus*	102
Eugenia caryophyllata	90	*Hevea brasiliensis*	102
Euonymus fimbriatus	90	*Hibiscus rosa-sinensis*	102
Euonymus grandiflorum	90	*Hibiscus tiliaceus*	103
Euphorbia ligularia	91	*Hiptage benghalensis*	103
Euphorbia prostrarta	91	*Holarrhena antidysenterica*	104
Evolvulus nummularius	91	*Holoptelea integrifolia*	104
Ficus altissima	92	*Hordeum vulgare*	105
Ficus hispida	92	*Hornstedtia arunachalensis*	105
Ficus racemosa	93	*Hydrocotyle javanica*	105
Ficus religiosa	93	*Hymenachne amplexicaulis*	106
Flacourtia ramontchi	94	*Ichnocarpus frutescens*	106
Flemingia bracteata	94	*Impatiens balsamina*	107
Flemingia chappar	95	*Impatiens gigantea*	107
Flueggea leucopyrus	95	*Indigofera astragalina*	107
Garcinia morella	95	*Indigofera cassioides*	108
Gardenia turgida	96	*Indigofera dosua*	108
Geranium himalayense	96	*Ipomoea aquatica*	108
Globba marantina	96	*Ipomoeapes-caprae*	109
Gloriosa superba	97	*Ixora barbata*	109
Gnetum scandens	97	*Ixora brunnescens*	110

Contd..

Appendix–XVII–*Contd...*

Contd..

Appendix–XVII–*Contd...*

Plant Species	*Page No.*	*Plant Species*	*Page No.*
Neolitsea chinensis	136	*Pogostemon heyneanus*	149
Nerium indicum	136	*Pongamia pinnata*	149
Nicotiana tabaccum	137	*Porana paniculata*	150
Nyctanthes arbortristis	137	*Potentilla leschenaultiana*	150
Nymphaea nouchali	138	*Pothos scandens*	150
Nypa fruticans	138	*Premna serratifolia*	150
Ocimum basilicum	138	*Prunus cerasoides*	151
Ocimum sanctum	139	*Prunus mahaleb*	151
Opilia amentacea	139	*Psychotria calocarpa*	152
Oroxylum indicum	139	*Pterocarpus marsupium*	152
Oryza sativa	140	*Pterocarpus santalinus*	152
Osbeckia stellata	140	*Punica roxburghii*	153
Oxalis corniculata	141	*Quisqualis indica*	153
Paederia foetida	141	*Rhynchostylis retusa*	153
Pandanus andamanensium	142	*Ricinus communis*	154
Pandanus fascicularis	142	*Rosa clinophylla*	154
Papaver somniferum	143	*Rosa indica*	155
Passiflora caerulea	143	*Rubia cordifolia*	155
Pavetta tomentosa	143	*Rubus insignis*	156
Phoenix dactylifera	144	*Rubus paniculatus*	156
Phoenix sylvestris	144	*Rumex scutatus*	156
Phragmites karka	144	*Saccharum munja*	156
Phyla nodiflora	145	*Saccharum officinarum*	157
Phyllanthus emblica	145	*Saccharum spontaneum*	157
Phyllocephalum phyllolaenum	146	*Samadera indica*	157
Pinus kesiya	146	*Sambucus nigra*	158
Pinus wallichiana	146	*Santalum album*	158
Piper betle	147	*Sapindus emarginata*	159
Pithecelobium angulatum	147	*Saraca asoca*	159
Plumbago indica	147	*Schima wallichii*	160
Plumbago zeylanica	148	*Schrebera sweitenioides*	160

Contd..

Appendix–XVII–*Contd...*

Contd..

Appendix–XVII–*Contd...*

Appendix-XVIII
Index to English Names

Contd..

Appendix–XVIII–*Contd...*

Plant Names	*Page No.*	*Plant Names*	*Page No.*
Bengal walnut	18	Bowstring hemp	46
Bermuda grass	75	Brab tree	39
Betel leaf vine	147	Bread wheat	182
Betel nut	27	Bridal creeper	150
Betel pepper	147	Brinjal	165
Bhutan pine	146	Broom grass	178
Biacon weed	59	Broomjute sida	163
Bile tree	167	Bryony	81
Birds eye chilli	49	Bucks-beard	180
Bitter gourd	130	Bullet wood	129
Bitter wort	174	Bur weed	190
Black Babool	10	Butea gum	42
Black creeper	106	Butterfly pea	65
Black cutch	9	Button tree	24
Black liquorice	50	Calamus	13
Black mangrove	15	Calbash Gourd	116
Black myrobalan	177	Californian cinchona	170
Black oil plant	57	Camel's foot climber	35
Black purification plant	115	Camel's thistle	82
Bleeding heart	64	Camil	11
Bleeding heart glorybower	64	Camphire	117
Bleeding heart vine	64	Candlenut tree	18
Blister craper	50	Canker time table	174
Blood wort	11	Cannabis	48
Blow ball	174	Caoutchouc Tree	102
Blue-flowered glory tree	64	Caper bush	49
Bonavist bean	116	Caper plant	48
Bonduc nut	42	Cardamom	85
Bot tree	93	Carilla fruit	130
Bottle Gourd	116	Carpernter-grass	11
Bouquet grass	178	Castor oil plant	154

Contd..

Appendix–XVIII–*Contd...*

Contd..

Appendix–XVIII–*Contd...*

Contd..

Appendix–XVIII–*Contd...*

Contd..

Appendix–XVIII–*Contd...*

Contd..

Appendix–XVIII–*Contd...*

Contd..

Appendix–XVIII–*Contd...*

Contd..

Appendix–XVIII–*Contd...*

Contd..

Appendix–XVIII–*Contd...*

Contd..

Appendix–XVIII–*Contd...*

Contd..

Appendix–XVIII–*Contd...*

Appendix-XIX
Index to Hindi Names

Contd..

Appendix–XIX–*Contd...*

Contd..

Appendix–XIX–*Contd...*

Contd..

Appendix–XIX–*Contd...*

Plant Names	*Page No.*	*Plant Names*	*Page No.*
Botornebu	61	Chhota-gokhru	190
Brachta	183	Chhota-italkusa	118
Brahmadundi	27	Chhota-sundi	138
Bujjer bhang	137	Chhoto gokhru	180
Burragokhur	180	Chian	87
Canchra	143	Chichra	12
Caricara	143	Chichra	45
Chagul nodi1	166	Chilauni	160
Chakotra	61	Chilbil	104
Chakramardah	55	Chilla	53
Chakunda	55	Chilla	53
Chalmori	141	Chilli	53
Chaluba	112	Chilmil	104
Chameli	111	Chindar	71
Chameli	112	Chini buti	160
Chameli	112	Chirchira	12
Chameli	171	Chirmiti	8
Champ	128	Chita	148
Champa	113	Chitawar	148
Champa	128	Chiti	148
Champa methi	181	Chitra	148
Champac	128	Chitra	148
Champaca	128	Chob chini	164
Chandanbedu	59	Chota cirayita	86
Chandni	171	Chota-kulpha	180
Charas	48	Chot-hal-kusa	118
Chatiun	21	Choti-elachi	85
Chatwa	30	Chucka	73
Chaupatira	125	Chui-Mui	128
Chaval	140	Chuka tripati	141
Chelwa	103	Chundrus	185

Contd..

Appendix–XIX–*Contd...*

Contd..

Appendix–XIX–*Contd...*

Contd..

Appendix–XIX–*Contd...*

Contd..

Appendix–XIX–*Contd...*

Contd..

Appendix–XIX–*Contd...*

Contd..

Appendix–XIX–*Contd...*

Contd..

Appendix–XIX–*Contd...*

Contd..

Appendix–XIX–*Contd...*

Contd..

Appendix–XIX–*Contd...*

Contd..

Appendix–XIX–*Contd...*

Contd..

Appendix–XIX–*Contd...*

Contd..

Appendix–XIX–*Contd...*

Contd..

Appendix–XIX–*Contd...*

Appendix-XX
Index to Sanskrit Names

Contd..

Appendix–XX–*Contd...*

Contd..

Appendix–XX–*Contd...*

Contd..

Appendix–XX–*Contd...*

Contd..

Appendix–XX–*Contd...*

Contd..

Appendix–XX–*Contd...*

Contd..

Appendix–XX–*Contd...*

Contd..

Appendix–XX–*Contd...*

Contd..

Appendix–XX–*Contd...*

Contd..

Appendix–XX–*Contd...*

Plant Names	*Page No.*	*Plant Names*	*Page No.*
Karavellah	131	Khasa	143
Karavi	50	Khurjjuri	144
Karavira	136	Kichaka	121
Karhata	58	Kinihi	65
Kari	166	Kinkirata	10
Kari padam	84	Kiriti	176
Karira	48	Kokileshta	171
Karkotaka	157	Kola	191
Karnasphota	50	Kolaphala	175
Karparala	19	Koya	153
Karpasamu	98	Kramuka	27
Karunta	34	Krichhari	80
Karuntaka	34	Krimighna	88
Karvir	136	Krimighni	137
Kasalu	157	Krishanatvaka	177
Kasha	157	Krishashakha	48
Kashaya	24	Krishna-keli	130
Kasturi	7	Krishnakhya	38
Katakali	70	Krishnakutayata	189
Kathillaka	38	Krishnala	8
Katu1	128	Krishna-nimba	132
Katuparni	27	Krishna-surasu	115
Kaveri	73	Kritamalluka	54
Kesara	127	Krsnaguruh	26
Kesharaja	82	Krura	38
Ketaka	142	Ksharapatra	137
Khadira	9	Ksharashreshtha	42
Khaggara	157	Kshiranasha	169
Kharachhada	175	Kshiri	182
Kharapatraka	137	Kuberaksh	42
Kharjura	144	Kucandanah	14

Contd..

Appendix–XX–*Contd...*

Plant Names	*Page No.*	*Plant Names*	*Page No.*
Kukkura	65	Mada	52
Kukubha	176	Madahvi	103
Kumara	128	Madana	190
Kumara	70	Madanabhaphala	19
Kumbhi	50	Madanaghanta	40
Kumuda	138	Madanaghanti	40
Kundah	112	Madhuka	123
Kundura	185	Madhukarkati	61
Kurasa	84	Madhukshir	144
Kuravaka	117	Madhumalati	189
Kutaja	104	Madhuparni	175
Kutaja	185	Madhuparnika	71
Kutsitangah	25	Madhura	30
Laghu gokshura	180	Madhuramlaka	167
Laghushleshmataka	69	Madhushaka	34
Lajjahah	129	Madhutrina	157
Lajjahah	129	Magajihva	86
Lajjalu	129	Mahabhita	129
Lajjalu	129	Mahabhita	129
Lakshmi	73	Mahaddala	59
Langali	66	Mahadroksha	126
Langalika	97	Mahajambu	171
Lanka	49	Mahakanda	19
Lasuna	20	Mahakapittha	160
Latakasturika	7	Mahakapittha	70
Latakshiri	182	Mahamoda	112
Latika	7	Mahamundi	166
Laxmi	139	Mahanimba	126
Lingini	81	Mahapatra	175
Lodra	169	Mahaphala	121
Lohita	38	Mahatika	165

Contd..

Appendix–XX–*Contd...*

Contd..

Appendix–XX–*Contd...*

Contd..

Appendix–XX–*Contd...*

Contd..

Appendix–XX–*Contd...*

Contd..

Appendix–XX–*Contd...*

Contd..

Appendix–XX–*Contd...*

Contd..

Appendix–XX–*Contd...*

Contd..

Appendix–XX–*Contd...*

Contd..

Appendix–XX–*Contd...*

Contd..

Appendix–XX–*Contd...*

Abrus precatorius

Acacia catechu

Agave americana

Allium sativum

Acorus calamus

Bryophyllum pinnatum

Butea monosperma

Calotropis procera

Cannabis sativa

Carica papaya

Cassia fistula

Citrus medica

Cynodon dactylon

Datura innoxia

Eclipta prostrata

Erythrina indica

Hibiscus rosa chinensis

Impatiens balsamina

Jacaranda mimosifolia

Jasminum grandiflorum

Justicia adhatoda

Nerium indica

Mirabilis jalapa

Papaver somniferum

Passiflora caerulea

Ricinus communis

Saraca asoca

Taxus baccata

Tinospora cordifolia

Urtica dioica

Xanthium strumarium

www.ingramcontent.com/pod-product-compliance
Ingram Content Group UK Ltd.
Pitfield, Milton Keynes, MK11 3LW, UK
UKHW022214051026
15624UKWH00001B/156

9 789351 240730